TRANSACTION LEVEL MODELING WITH SYSTEMC

Transaction Level Modeling with SystemC

TLM Concepts and Applications for Embedded Systems

Edited by

FRANK GHENASSIA
ST Microelectronics, France

 Springer

A C.I.P. Catalogue record for this book is available from the Library of Congress.

ISBN 10 0-387-26232-6 (HB)
ISBN 13 978-0-387-26232-1 (HB)
ISBN 10 0-387-26233-4 (e-book)
ISBN 13 978-0-387-26233-4 (e-book)

Published by Springer,
P.O. Box 17, 3300 AA Dordrecht, The Netherlands.

www.springeronline.com

Printed on acid-free paper

Printed in the Netherlands.

Contents

Foreword

System-on-Chip and TLM

A System-on-Chip (SoC) is a blend of software and silicon hardware components intended to perform a pre-defined set of functions in order to serve a given market. Examples are SoCs for cell phones, DVD players, ADSL line cards or WLAN transceivers. These functions have to be delivered to the target users as a SoC product during the right market window at satisfactory levels of performance and cost.

Over the past 20 years, the productivity of SoC designers has not been able to keep pace with Moore's Law, which states that the silicon process technology allows doubling the number of transistors per chip every 18 or 24 months. Since the advent of RTL, designers and design automation engineers have searched for the next design methodology allowing a step function in design productivity.

Simply put, we believe that we have found and delivered to the industry the next SoC design methodology breakthrough: *System-C TLM*. This book is a vibrant testimony by the people who made it happen, giving both some details on the search for this Holy Grail, and the many facets of the applications of TLM.

The Search for SystemC TLM

Raising the level of CMOS digital design abstraction from gate-level and schematic capture to Register-Transfer-Level (RTL) has enabled a fundamental breakthrough in digital circuit design in the 1980s and 1990s. RTL's clean separation between Boolean operations on signals, and clocks registering the results of these operations, was first embodied in the Verilog language initially designed by Phil Moorby in 1985; then in VHDL with the initial IEEE standard approved in 1987. RTL was first thought of as a more efficient way to model digital designs. Soon, its wonderful formal characteristics allowed separating combinatorial logic optimization as demonstrated by MIS[1], from sequential elements such as registers or latches. In turn, complete synthesis tools emerged, as exemplified by Design Compiler from Synopsys.

Since RTL, many attempts have been made at identifying and defining the 'next' practical level of design abstraction. Of course, algorithm developers start out at a very abstract level, which is not tied to any architecture decision or implementation. What missing was an intermediate level, which would be abstract enough to allow complete system architecture definition while being accurate enough to allow performance analysis.

In 1999, a small motivated team of researchers from various fields at ST set out to design and verify a third-generation H263 video CODEC[2], architectured with several dedicated heterogeneous processors as well as several hardware accelerators. As other SoC architects, they had to identify performance bottlenecks of the CODEC, while simultaneously defining and refining the micro-architecture of the hardware accelerators, the instruction set of the dedicated processors, and the embedded software performing control tasks and handshaking with the external world. On a previous incarnation of the CODEC, the designers had used extensive RTL-based verification methods, including hardware emulators, in order to verify the embedded software running on the selected micro-architecture with hundreds of reference image streams.

Every time a functional or performance issue requiring an architecture or micro-architecture change was encountered, a long re-design and re-verification cycle, spanning many weeks and sometimes months, would be necessary.

On the other hand, for the embedded software developer working with the processor architect, a modification requiring a change of the instruction set was almost immediate: a new Instruction Set Simulator (ISS) was generated and the embedded software could run very rapidly on the new ISS. The reason was that the processor was modeled in C as a functional model, and some wrapper code that represented the interface and communication to the processor peripherals.

During a project review the idea emerged that, using the same abstraction level as the ISS for other SOC hardware blocks would allow a breakthrough

---

[1]   R. Brayton, R. Rudell, A. Sangiovanni-Vincentelli, and A. Wang. MIS: A multiple-level logic optimization system. IEEE Transactions on Computer-Aided Design of Integrated Circuits and Systems, CAD-6 (6), Nov. 1987

[2]   M Harrand *et al*, "A single-chip CIF 30-Hz, H261, H263, and H263+ video encoder-decoder with embedded display controller", IEEE journal of Solid State circuits, Vol 34, No 11, Nov 1999

in verification time. In itself, the idea of dissociating cleanly function and communication was not new, but the real breakthrough came from developing a framework for this modeling abstraction using an open and still evolving design modeling language: SystemC.

Using SystemC as a vehicle to provide the Transaction Level Modeling (TLM) abstraction proved to be the key to the fairly fast deployment of this methodology. There was no issue of proprietary language support by only one CAD vendor or university. There was also no issue of making a purchase decision by the design manager for yet another costly design tool.

Eventually, with the collaboration of ARM and Cadence Design Systems, a full-blown proposal was made to the Open SystemC Initiative (OSCI), under the name PV (Programmer View) and PVT (Programmer View Timed). Indeed 'Programmer View' clearly reflects the intent of this new abstraction level, which is to bridge the gap between the embedded software developer and the hardware architect.

Paradigm Shift

Not all the possible implications of sharing a single executable functional reference across the various teams have been explored yet.

Certainly, allowing the Algorithm, Hardware, Software and Functional Verification teams to rely on the same functional model is saving valuable time by avoiding misunderstandings due to informal or even formal paper-based communication.

However, we are also witnessing a real paradigm shift in the way software and hardware engineers work with each other. When an SD video movie can run at the rate of 1 image/second, equivalent to 12MHz, on an early model of the architecture, this allows SW development to start while the architecture is not yet frozen. Of course, earlier interactions between the hardware and software teams lead to better overall SoCs. Since more and more, delivering a prototype to the SOC customer is on the critical path of the application software development by that customer, TLM-based SoC platforms actually allow early application software development by the end customer before the actual hardware architecture is even frozen.

Next, a full ecosystem of system-level IP developers, both in-house and from third-party vendors, needs to develop. We are taking steps in raising the awareness level of the IP providers, so they start to include these TLM views as a standard part of their deliverables together with RTL models. Beyond this, we are making fast progress within the SPIRIT consortium, which will allow the SoC architect to mix and match IP blocks modeled in TLM, as system-level IP functional descriptions.

Philippe Magarshack
Crolles, April 18th 2005

Preface

Throughout the evolution of microelectronics industry, SoC designers have always been struggling to improve their productivity in order to fully exploit the growing number of transistors on a chip achievable by the silicon process capacity.

The answer to this challenge has always been increasing the level of abstraction used for the SoC implementation. From transistors to gates, and from gates to RTL, the design productivity has been maintained high enough to keep pace with and take advantage of the silicon technologies. Unfortunately, RTL as the design entry point cannot handle the complexity of 500 million-transistor SoCs designed with the CMOS90 process technology.

Two major directions are contributing to bridge the gap between design productivity and process capacity:
- Raise the level of abstraction to specify and model a SoC design.
- Adopt a different design paradigm, going from hardwired blocks to partially or fully programmable solutions, as pioneered by Paulin et al[1].

Transaction Level Modeling with SystemC presents an industry-proven approach to address the first direction. The proposed solution resolves critical system level issues encountered in designing a SoC and its associated embedded software. The brief history of our reaching TLM at STMicroelectronics is traced in Chapter 1.

---

[1] P. G. Paulin, C. Pilkington, M. Langevin, E. Bensoudane, and G. Nicolescu, "Parallel Programming Models for a Multi-Processor SoC Platform Applied to High-Speed Traffic Management," in Proc. of International Conference on Hardware/Software Codesign and System Synthesis (CODES+ISSS), 2004 (Best Paper Award).

TLM, an acronym for Transaction Level Modeling, has become an overloaded buzzword hiding too many different abstractions and modeling techniques. Applications of our TLM definition as described in Chapters 2 and 3, have proved to successfully tackle the following topics:

- *Productivity* through a veritable hardware/software co-development based on virtual prototypes, as described in Chapter 4.
- *First-time Silicon Success* (FTSS) achieved by using TLM as golden reference in the functional verification flow, which also enables a system-oriented verification, as described in Chapter 5. Ensuring the compliance of the SoC design with real-time constraints of the targeted application also contributes to FTSS, as discussed in Chapter 6.
- *Efficient workflow* between the numerous teams contributing to the development of the SoC and associated software. This is attainable by sharing a unique set of specification documents and models, as well as by keeping consistency between the various teams through platform automation tools, as described in Chapter 7.

This book is intended for engineers and managers who face challenges of designing SoCs in advanced CMOS technologies, and seek for solutions to enhance their current SoC and system level methodologies. It also serves engineers looking for SystemC modeling guidelines. More generally, we hope that this book will trigger new ideas in the research community to enhance design techniques based on Transaction Level Modeling.

Acknowledgements
This book is the result of five years of research and development work at STMicroelectronics. The authors of the chapters are the main architects of the resulting environment. I wish to thank all of them here for their key contributions. Our solution could not be industrialized without the great help of many engineers, especially knowing that "the devil is in the details" as the old saying goes. I am grateful to all of them for their great contributions: Aditya Raghunath, Adnene Ben-Halima, Alain Kaufman, Amandeep Khurmi, Amit Mangla, Andrei Danes, Ankur Sareen, Anmol Sethy, Arnaud Richard, Ashish Trambadia, Bruno Galilée, Christophe Leclerc, Claude Helmstetter, Dinesh Kumar Garg, Dheeraj Kaushik, Dorsaf Fayech, Emmanuel Viaud, Etienne Lantreibecq, Hervé Broquin, Jérôme Cornet, Jérôme Peillat, Julien Thevenon, Kamlesh Pathak, Kshitiz Jain, Laurent Bernard, Maher Hechkel, Mamta Bansal, Maxime Fiandino, Michel Bruant, Mukesh Chopra, Naveen Khajuria, Olivier Raoult, Ramesh Mishra, Rohit Jindal, Sandeep Khurana, Stephane Maulet, Tran Nguyen, Vincent Motel, and Walid Joulak.

Many from the industry and the academia have guided us in the right direction. Special thanks go to Ahmed Jerraya, Chuck Pilkington, Florence

Maraninchi, Jean-Michel Fernandez, John Pierce, Joseph Borel, Laurent Ducousso, Marcello Coppola, Mark Burton, Michel Favre, Pierre Paulin, and Thorsten Grötker.

I also appreciate our employer for giving us ample freedom to propose and experiment new ideas, especially Philippe Magarshack, for his continuous support and encouragement.

Last but not least, I wish to acknowledge the special contribution of Jim-Moi Wong. Writing a book while employed in the industry is a great challenge because other urgent tasks consume most of the time. Despite such situation, her professionalism and dedication have smoothly implemented this book project; and finally, we are able to publish this book.

Frank Ghenassia
Crolles, May 2005

Special Contribution

All of the material and technical information in this book are collected, organized, elaborated, structured, and drafted by Jim-Moi Wong through the very tight cooperation with the editor-in-chief, the co-authors, and the team of System Platform Group (SPG) at STMicroelectronics.

Chapter 1

TLM: AN OVERVIEW AND BRIEF HISTORY

Frank Ghenassia and Alain Clouard
STMicroelectronics, France

Abstract: The trend of "the smaller the better" in semiconductor industry pictures a bright future for System-on-Chip (SoC). The full exploitation of new silicon capabilities, however, is limited by the tremendous SoC design complexity to be addressed within very short project schedule. This limiting factor has pushed the need for altering the classic SoC design flow into prominence. A novel SoC design flow starting from a higher abstraction level than RTL, i.e. System-to-RTL design flow, has surfaced as a real need in advanced SoC design teams. After a decade of attempts to define a useful intermediate abstraction between SoC paper specification and synthesizable RTL, the SystemC C++ open-source class library has finally emerged as the right vehicle to explore the adequate level of abstraction. Transaction Level Modeling (TLM), a methodology based upon such abstraction, has proven revolutionary values in bringing software and hardware teams together using the unique reference model; resulting in dramatic reduction of time-to-market and improvement of SoC design quality.

Key words: system-on-chip; integrated circuit; SoC bottleneck; system-to-RTL design flow; transaction level modeling; TLM; abstraction level; SystemC; OSCI.

1.   SYSTEM-ON-CHIP

1.1   The Smaller The Better

An electronic system is a blend of hardware and software components intended for performing a set of functions. These functions have to be delivered to target users at a satisfactory level of performance.

The integrated circuit (IC) or chip is a semiconductor wafer comprising millions of interconnected transistors as well as passive components such as resistors and capacitors. ICs can function as any individual or combined

F. Ghenassia (ed.), Transaction Level Modeling with SystemC, 1-22.

parts of an electronic system, for instance, microprocessors, memories, amplifiers, or oscillators. In general, ICs are classified into three categories according to their intended purposes: analog, digital, and mixed-signal.

Through the tiny size of a few square millimeters, integrated circuits have dramatically improved the overall system performance compared to those circuits assembled at board level. High speed, low power consumption, and reduced fabrication cost are among the most remarkable benefits brought by ICs.

In 1965, Gordon Moore predicted that the number of transistors incorporated in an IC would increase twofold every year. This was really an amazing prediction proved to be more accurate than Moore had believed. Since the past few decades, the scale of IC integration has been soaring high. It started from Small Scale Integration (SSI) with around 100 transistors per IC in 1960s, up to Very Large Scale Integration (VLSI) accommodating more than 10000 transistors per IC in 1980s. There is no sign that such tendency would ever cease. In recent years, the integration scale has only slightly slowed down to a factor of two for every eighteen months. This very interesting observation has later been adopted by the Semiconductor Industry Association (SIA) as the famous Moore's Law to determine IC complexity growth.

Nowadays, ICs could hardly be removed from daily life since they are extensively used in consumer electronic products, telecommunication, data processing, computing, automotive merchandises, multimedia, aerospace, industry and so forth. This invention has really made great changes in our modern life style. Integrated circuits are, for this reason, widely acclaimed as one of the most important inventions in the last century.

The outburst of IC complexity, as predicted by Moore's Law, is driving the current semiconductor industry to challenge another cutting edge revolution: *System-on-Chip (SoC)*. With the capacity of integrating more and more transistors in a chip, the principle of "the smaller the better" seems steadily realistic and promising.

System-on-Chip is the concept of conceiving and integrating distinct electronic components on a single chip to form an entire electronic system. This concept is feasible thanks to the very exceptional manufacturing advances that bring IC nanotechnology to fruition.

SoC is typically used in a small yet complex consumer electronic product such as hand-phone or digital camera. The fundamental building blocks of SoC are intellectual property (IP) cores, which are reusable hardware blocks designed to perform a particular task of a given component. An IP core could either be a programmable component like processor; or a hardware entity with fixed behavior like memory, input/output peripheral, radio-frequency analog device, and timer. The different IP cores are interconnected on a SoC by some communication structures such as shared buses or network-on-chip (NoC), in order to establish communication among them.

A very frequent practice today is to group IP cores and communication structures on a so-called *SoC platform* to create an application-specific SoC template. Such platforms provide users with ample room for product differentiation at reduced design time and effort; and thus the final SoC product can be delivered in a timely manner to market. With the advent of the latest complementary metal oxide semiconductor (CMOS) technologies, a SoC platform comprises not only hardwired functions but also the embedded software. More often than not, the embedded software runs on multi-processors that all present on the same SoC.

System-on-Chip has brought to mankind a new field of boundless imagination. Through its tiny little size empowered with high performance as a whole system, SoC is undoubtedly a major breakthrough in the semiconductor industry. Imagine, a blind child could probably be able to see the bright world again thanks to a tiny bio-electrical chip implanted in his or her brain; or wearing a hand-held personal computer on your wrist could be as common as wearing a watch very soon!

Yes, the future of the semiconductor and consumer industries relies heavily on SoC. When considerations are given to all the complex factors constituting a SoC, however, plenty of challenges would simply start to accumulate right in front of us: "How do we manage the intricacy of SoC design procedures yet sustaining a satisfactory product quality?"

To better analyze this subject, the role of the classic SoC design flow must first be identified, followed by an examination of how the current SoC bottlenecks have limited its performance and what could be done to tear these barriers down.

1.2   Classic Design Flow

The design flow is a rigorous engineering methodology or process for conceiving, verifying, validating, and delivering a final integrated circuit design to production, at a meticulously controlled level of quality.

Traditionally, digital electronic embedded systems employ the design flow as illustrated in Figure 1-1. Such flows set off from a general picture of the system specification. It then splits into two distinct paths of activities:

1. system hardware development;
2. system software development.

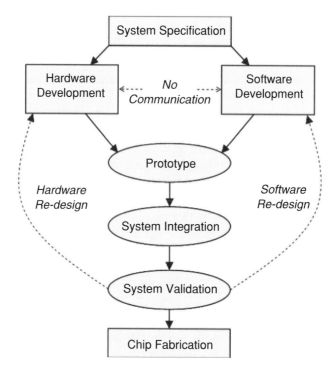

Figure 1-1. Classic Design Flow

Note that there is *no* communication at all between these two paths of design work. The hardware and software design can only be conducted separately until a prototype of the system-under-design is made available.

To understand this classic design flow, let us begin with the hardware path. Here, the job starts rolling from the register transfer level (RTL) code development. This step is accomplished by creating hardware models using hardware description language (HDL) such as VHDL or Verilog. These models will go through functional verification in simulation to attest the correctness of their behavior. Subsequently, synthesis is performed to obtain a logic netlist. The hardware design has so far gone through the front-end logic design steps. Once the netlist is ready, it will enter the back-end design steps; typically ranging from layout drawing to floorplan, place and route, resistance and capacitance extraction, timing analysis, and all the way down to physical verification. Now, the hardware design is essentially a tape-out version readily being sent to fabrication for building a prototype of the system.

On the software path of the classic design flow, the system embedded software will be developed independently of the hardware design. Software engineers will just sit in their own corner and write up the software codes without thinking that they may need to talk to hardware engineers. Although the coding could be started soon this way, testing the software accessing new peripheral IPs requires mapping RTL codes on an emulator or FPGA-based prototype system. This is a costly process involving expensive equipment. A worse situation is waiting for the test chip from the fab in order to test it on a prototype board. As a result, the software is always validated later than the hardware.

Once the system prototype is available, the software will be embedded into the prototype to conduct system integration and validation. If any errors are found in the hardware or software, the design process will be iterated as indicated in Figure 1-1. These loops might repeat until a good functioning system with adequate performance is attained. Finally, the design is sent to the fab for volume production.

During the 1990s, the so-called "co-verification" was used to jointly simulate the RTL hardware and embedded software [1]. However, it was running at the slow speed of RTL, typically hundreds of bus cycles per second for a complete SoC with only one or two processors and a dozen of mid-size peripherals. Thus, the co-verification could only run small software codes, for instance, debugging software drivers of simple devices. Since software applications are getting way too complex under a constantly shrinking time-to-market, the co-verification could not cope with the situation. What the SoC industry needs now is a hardware/software co-simulation that can simulate the hardware at higher speed.

1.3   SoC Bottlenecks

The vigorous trend of decreasing the minimum feature size on an increasing wafer dimension is almost a point of no return when SIA Roadmap traces the forecast of Moore's Law. This exponential tendency is pushing the contemporary SoC era to challenge its peak [2-3]. Such challenge can be sorted into three major bottlenecks as follows:

- **Explosive Complexity**

  A rather troubling dilemma is the complexity that comes along with the ground-breaking SoC evolution. While SoC industry struggles for its ultimate goal of "the smaller the better", more and more functions are incorporated into a system to perform increasingly sophisticated tasks.

  A typical SoC integrates many blocks including peripheral IPs, buses, complex interconnects, multiple processors (often of different kinds), memory cuts, etc. There are always several master blocks on the bus or interconnect, resulting in complex arbitration of communications and difficult estimations of bandwidth and latency. The complexity for those SoCs under new design or planned for the next generation can easily exceed the complexity of current SoCs.

The tricky game of SoC design does not simply deal with the flawless multifaceted-team cooperation to produce a complete SoC ranging from design to process. The direct impact on the overall SoC performance must also be carefully handled throughout the whole design cycle. Rigorous methodology must be implemented to address reliability issues of not only how a SoC performs, but also of how *good* a SoC can perform reliability issues.

Given such complexity, the reliability of SoC performance must be assured accordingly starting from earlier, higher, and stricter level. This is unfortunately a very tough and time-consuming job to cope with. Not even the slightest error should be tolerated because that will simply snowball the problem with increasing correction costs as the design advances.

The reliability reinforcement must span widely throughout the entire SoC design and process flow. This methodology should tackle every design and process level that could have an impact on the overall SoC performance, i.e. verification, validation, integration, timing and power checking, chip testing, and packaging.

An additional factor making new SoCs more difficult to design is the type of software applications running on their processors. Consumers can now purchase electronic products with multiple capabilities. For example, a modern hand-phone has to embed MP3 player, radio, PDA functionalities, and digital camera in addition to its basic functions of handling incoming voice or video calls.

The architecture study using standard methods such as spreadsheet formulas or point simulations (with critical software benchmarks running on RTL model of limited hardware), can result in over-dimensioned buses, processors, memories, etc, due to margins introduced by uncertainties. Sadly, the over-dimensioned SoC architecture will only lead to a non-competitive silicon area. The architect requires a fast yet accurate simulation of the complex SoC running the real application software (at least a significant part of it).

The current SoC design can no longer survive on the traditional design flow considering all the complexity factors. Instead of the classical approach where separate teams work on various incoherent models, what the SoC design really needs now is an expanded space that links all the different phases of the design through a centralized methodology.

- **Time-to-Market Pressure**

Time-to-market is the amount of time required for conceiving an idea into a real product for sales. Every product has a market window. If time-to-market is shortened, the product will be available earlier in the market for gaining larger market share and earning higher revenue. For certain markets, the first product still occupies about 60% of the market share even after the competitors have offered the alternative products.

Today, the fast-moving market does not allow superfluous time loss in product development and production; you may otherwise pay the dear price of missing the market window. A typical example is delivering consumer products by a particular date of some special festivals.

The increasing complexity of current SoC products usually necessitates time-consuming development phases. This has critically hindered the attempt to shrink the time-to-market of SoC products. The classic design flow is unfortunately of little help in this case because it is always too long to wait for a prototype. Instead, a more flexible and efficient methodology is sought after to optimize the time management of SoC projects.

- **Sky-rocketing Cost**

The ever-increasing cost of SoC development and production, close to an unacceptable level lately, is probably one of the most nerve-racking worries of SoC industry. Since the current SoC design necessitates higher workforce and much costly masks, re-spins due to errors in design functionality or performance are not tolerated, i.e. first-time system success is critical.

Due to the tremendous complexity of the current SoC products, a larger workforce must be provided to manage tricky problems encountered in design, verification, and manufacturing. In parallel with the growing complexity, Electronic Design Automation (EDA) tools intended for designing and verification are getting much more expensive. On top of it, the hottest spotlight of SoC production -*nanotechnology*- has dramatically raised the costs of manufacturing equipment and facility greater than ever.

Here again the dilemma: "How could SoC industry sustain this unreasonable cost burden while trying to keep up with the projection of Moore's Law?"

Not surprisingly, the traditional design flow cannot do much good in solving this problem. Some of the semiconductor foundries start to form alliances to share the production cost based on the same manufacturing technology. More fundamentally, revolutionary methodology approaches to design and verification should be phased in to strike at the roots of cost issues.

2. SYSTEM-TO-RTL DESIGN FLOW

2.1 The Need for a Novel Design Flow

SoC bottlenecks have propelled the whole SoC industry to ponder on its future seriously. Countless discussions and researches have been going on since years to hunt for the most favorable solution.

IP reuse is one of the important research directions. Nonetheless, it has some drawbacks. The time spent to identify, understand, select, and integrate a third-party IP places this approach at an unfavorable position compared to designing it in-house. The situation could be worse if the IP provider does not react promptly for any integration issues encountered.

Another prevailing direction pursued is raising the design abstraction above the register transfer level (RTL), an approach generally known as *system level design*. This approach adds an extension of *system-to-RTL* design flow on top of the standard RTL-to-layout design flow so that the entry point of SoC design resides at higher abstraction level than RTL.

Many good reasons make it convincing to extend the classic design flow to system level. First of all, think over the importance of shortening time-to-market of SoC products. Due to the explosive complexity, the software must provide a considerable part of the expected SoC functions to alleviate lengthy hardware design process. It grants flexibility for product evolutions either during the design with evolving standards, or during the deployment with field upgrades. For example, downloadable video CODEC update or user-selected applications download for hand-phone games. Strategically, the system-to-RTL design flow enables developing and testing the software earlier to accelerate the SoC design cycle.

Second, it is believed that system level design has promising potential to well perform architecture analysis and functional verification. These are crucial issues in SoC development today. Analyzing the expected real-time behavior of a defined SoC architecture could be very critical since real-time requirements are key specification parameters for many SoC targeted application domains, for instance, telecommunication, multimedia, or automotive. System level simulation and analysis is the right initial flow step to handle the difficult issue of not over-dimensioning the SoC hardware architecture that runs dynamic application software.

The third reason of extending the design flow to system level is for hardware design verification, which accounts for about 50 to 70% of SoC project effort. Good SoC design flows should support efficient verification processes for attesting SoC functional behavior and performance resulting from system integration of IPs. Efficient verifications reduce not only SoC development time but also the risk of the dreadful silicon re-spin.

2.2   Brief History of Our Reaching TLM

Any design and verification flow requires some defined abstraction level for models on which the flow tools can operate. To start SoC architecture, design, and verification from a higher level than RTL, the right type of abstracted modeling must be identified to support system design activities for both hardware and software engineers. Bear in mind that having more model views means involving higher development cost and complex management for the coherence among the different views.

2.2.1   Efforts on Cycle-Accurate Modeling

In late 1990s, many large companies started to develop their own models while research institutes and EDA start-ups were proposing a variety of modeling languages. Among the proposed languages, some of them were built from scratch. Some were "extended subsets" of the existing general-purpose software programming languages especially those of object-oriented languages such as C++ or Java. The examples include SpecC [4], CowareC, and VCC classes. Other proposed languages were extensions of hardware HDLs such as VHDL or Verilog. A typical example is Superlog. ICL had an interesting multi-level modeling approach for systems but we were not seeing them an as an EDA tool supplier.

As a central system flow team, we developed different kinds of models for various SoC projects using several of these languages. The models were developed at various abstraction levels depending on the requests from various SoC design teams in the company. The initial requests were to have cycle-accurate C or C++ models from certain who believed that it was the right way to get simulations running at least one order of magnitude faster than RTL models in VHDL or Verilog. It soon became obvious that cycle-accurate modeling had several drawbacks.

First, the modeling effort was close to the one of creating synthesizable RTL models. It was due to fact that the model complexity was too close to RTL. The only gain was that such models had no synthesis-related constraints. In addition, the RTL was still the reference due to immature synthesis tools. It led to iterations of the C++ model trying to keep in line with the RTL model of the IP under design. Introducing any specification change in the C++ model during the design was almost as long as doing so in the RTL model. The cycle-accurate modeling was actually leading to high costs. These models were not available to architects and were ready for software developers a little too late. Second, the simulation speed for a SoC model was ten times below the original objective. It was simulating at a few kHz compared to the several hundreds of Hz for RTL. Third, using specific languages or modeling optimizations to gain speed was actually locking the modeling team into a specific simulator supplier. Fourth, during final RTL updates before tape-out, it was usually not possible to keep updating the cycle-accurate C++ model due to tight schedule. Thus, the cycle-accurate model was not fully consistent with the reference RTL at tape-out. Normally, modeling engineers would be allocated to another project once the SoC was taped-out. The model would not be usable as a starting point for its next generation design because it was not consistent with the existing RTL and original modeling engineers were unavailable.

For all these reasons, we were looking for an higher level of abstraction that would allow much quicker modeling than cycle-accuracy, yet be precise and fast enough for software developers to test the real embedded software using a standard language enabling reuse of models with a variety of simulator suppliers. Ideally, such models should also be usable for performance estimations with enough precision for SoC architects to make decisions.

2.2.2   Our Road to SystemC-based TLM

In 1999, two of our suppliers, Coware and Synopsys, came to us with a proposal to support the standardization of a C++ set of classes for hardware modeling. This proposal was made with an open-source reference simulator that was to be completed by commercial refined features, commercial simulators, and other system tools for architects. We considered this initiative as the first real step addressing the need for system language standardization and the model reuse across various tools from a future market of EDA system tool suppliers. Hence, we decided company-wide to support SystemC as the language to be used as the basis for our efforts of defining an appropriate system modeling methodology.

SystemC 0.9 included RTL constructs but also some initial channel concepts that could be analyzed as the right direction for more abstract modeling than RTL. However, SystemC 1.0 was lacking of such high-level channels and was totally targeting RTL, i.e. cycle-accurate type of modeling, as we had already practiced with other languages.

We made serious moves for the SystemC issue. At OSCI SystemC steering group meeting, Alain Clouard presented requests for more abstract concepts, in particular to support modeling driven by system specification events but not design implementation clocks. SystemC 2.0 was then specified by OSCI language working group with system-level constructs such as new channels as well as inputs from colleagues especially Marcello

Coppola, from an earlier STMicroelectronics C++ modeling library named IPsim.

For the rest of 2000, we continued to work in parallel on more abstract modeling than RTL using other languages enabling such methodology. Using Cadence VCC, Giorgio Mastrorocco upgraded a Parades model [5] of a dual processor SoC of STMicroelectronics. Our team, in partnership with Cadence, compared its performance estimates precision against RTL [6] (which received best paper award of DATE'2002 Industry Forum).

We further refined our plan according to our requirements, for instance, a simulator for cycle-less models based on SoC specification events and managing time without cycles for fast simulation. Although lighter to code than RTL, transaction level modeling would require an initial investment in creating a library of commodity IP models. This was essential to adopt TLM as a methodology. It was also clear that SoC models with multiple masters on the interconnect would need to really execute read/write transactions from the RTL and change values in memories and peripheral registers. A performance model was simply too complex to build for an architect-only usage, and could not be used by software engineers to perform functional testing of their embedded software.

The real proof of modeling efficiency at transaction level came early 2001 when, using Unix IPCs, Etienne Lantreibecq and Laurent Maillet-Contoz enhanced a high-level C behavioral model from Joseph Bulone and Jose Sanches of an ST H263 video CODEC multiprocessor macro cell, to create a speed-efficient, bit-accurate, cycle-less, concurrent multi-IP modeling of the macro cell in only a few weeks. The model was running orders of magnitude faster than RTL, and updated before the RTL update for MPEG4; hence allowed to develop concurrently and efficiently the embedded firmware, dimension the code memories sizes, and sign-off the architecture with granted competitive silicon area for MPEG4 macro cell.

Once SystemC 2.0 appeared as early release on July 13th 2001, we immediately started to evaluate higher-level features of the language, e.g. new channels, events, and achievable simulation speed, by creating simple transaction level models with the key SoC architectural concepts. Jean-Philippe Strassen, with the experience from earlier efforts described above, developed a first SystemC 2.0 model of SoC showing implementation of TLM abstractions for the main components of an SoC: bus model including address management, bus master (one or multiple instances) creating read/write transfers, memory, timer, and interrupt controller with a thread in the bus master handling the interrupts. Our initial SystemC 2.0 TLM platform simulation without any optimization was around thousand times faster than the equivalent RTL or cycle-accurate C models simulations.

Combining the facts that the IP modeling effort was much less than the one required for modeling at bus cycle-accurate (BCA) or RTL level, and that the simulation speed was fast enough, we decided to continue the investigation of SystemC-based transaction level modeling.

We implemented the canonical SoC platform in various flavors of modeling on top of SystemC 2.0, such as bidirectional TLM `transport()`[1] call or unidirectional transfers such as put and get calls [7]. Models were 30% slower with the put/get approach compared to the transport approach in simulations on the RISC workstations. This made a real difference with respect to the speed-up that we were targeting for TLM. Therefore, we decided to adopt the bidirectional approach for our methodology, i.e. `transport()`.

To compare SystemC 2.0 relevance for TLM modeling effort and simulation speed, we also implemented the same canonical simple SoC in other languages including Unix IPCs scheme of the H263 work.

Among them, SystemC was the most flexible approach for modeling inter-IP communications and synchronization. It enabled exploiting the speed of C/C++ models for the internal behavior, which represented the majority of the simulation time expense for a real SOC, compared to communications and synchronizations. Further, SystemC was the only proposal for standardization with tool roadmaps from commercial vendors, and an open-source simulator facilitating the adoption of the new TLM abstract view.

Our canonical simple SoC TLM platform was a key demonstrator for all the optimizations in our TLM base classes for improved methodology and faster simulation speed. The H263/MPEG4 CODEC TLM model was running at a similar speed compared to running the RTL model of the same design on the most costly emulator: 2.5 seconds for coding and decoding an MPEG4 image. The VHDL RTL simulation was taking one hour. The TLM methodology was presented at FDL (Forum of Design Languages) by Frank Ghenassia in Marseille, France.

In 2001, our team reached several milestones through SystemC simulation. We worked with Cadence on the joint specification describing how our SystemC model could work with the VHDL model of another IP as mixed-language simulation, which was then implemented as prototype simulator by Cadence.

By Mid 2002, we had developed about twenty TLM IP models used as main subset of SoCs. The first session of our 5-day SystemC and TLM training was held for STMicroelectronics engineers. In 2002, we obtained

---

[1] Included in 2005 OSCI TLM standard.

the first benefit of SystemC TLM models for STMicroelectronics products. Kshitiz Jain and Rohit Jindal developed a SoC TLM model, enabling a four-month gain for one of our divisions starting embedded software development earlier than the availability of RTL on FPGA fast prototyping system. It yielded an embedded boot loader software fully functional and unchanged when run first time on RTL. We progressed steadily on optimizations for TLM simulation speed with contributions from Pierre Paulin, Chuck Pilkington and colleagues from STMicroelectronics Ottawa [8]. Our team was also progressing towards the idea of standardizing TLM both internal of STMicroelectronics and in OSCI. Meanwhile, others were also getting the benefits of TLM modeling using SystemC, e.g. the OCP-IP regarding the abstraction levels [9].

It was important to have hardware teams to use TLM. Towards end of 2002, we successfully beta-tested Cadence mixed-language SystemC/HDL simulator after the joint 2001 prototype of STMicroelectronics and Cadence was made. Based on Antoine Perrin and Rohit Jindal work, we demonstrated our canonical SoC platform running half of the models in VHDL RTL and half in SystemC TLM to ST divisions.

Since early 2003, TLM was widely deployed in STMicroelectronics not only for software development but also as reference models in functional verification for RTL IPs. We started to see cross-functional teams exchanging IP models, breaking the wall between hardware and software engineers, and spotting issues in paper specifications and inconsistencies between parallel development of software and RTL. Divisions observed gains in simplicity of environment setup and simulation speed of SystemC models in their functional verification test bench compared to their earlier approaches. In 2004, we reached several hundreds kHz execution of a SoC half on workstation TLM simulation and half on an FPGA board (much less costly than an emulator), thanks to our synthesizable STBus adaptors between TLM and RTL developed by Mukesh Chopra and colleagues, and based on the SCE-MI standard for transaction-based co-emulation [10].

On the speed side, Serge Hustin and colleagues further demonstrated in 2003 the power of TLM, leveraging their own methodology experience on abstract system models, by creating in a few weeks a SoC simulation of the core of a modem that was simulating on a workstation at a third of the speed of the actual chip.

After two years working on TLM modeling, we were invited to contribute a chapter for SystemC methodologies and application [11] in 2003. In the same year, TLM was standardized in the STMicroelectronics hardware design rules standard manual, the BlueBook. Meanwhile, we proposed together with Cadence and ARM a foundation proposal to OSCI for the new OSCI TLM working group. With this OSCI TLM WG that involved more companies, e.g. Philips and Mentor Graphics, the board of the Open SystemC Initiative approved the TLM standard on April 21st 2005.

The deployment of SystemC TLM for functional testing of embedded software and hardware RTL entails using TLM IP models in SoC integration. Some algorithm teams also use TLM as a way to structure their algorithm developments and make them readily usable for other teams. We have noticed that architects who are typical senior experts with both HDL and C++ knowledge have yet to exploit TLM benefits for their architecture studies. However, advances described later in this chapter about TLM modeling with time annotations (timed TLM or PVT), RTOS emulation with native compilation scheduling on top of TLM platforms, further usage of TLM before hardware/software partitioning, and model transformation techniques from standard specifications such as UML, are all helping to get TLM profitable for SoC architects. The remainder of the chapter introduces the main concepts and the next advanced usages of TLM that make it a powerful abstraction for SoC projects.

3.   TRANSACTION LEVEL MODELING

Through experiences and results gained from our tireless research and development, we propose a bit-true, address-map accurate, cycle-less *Transaction Level Modeling (TLM)* based on events from the hardware/software system specification as a sound solution to system level design.

TLM is a transaction-based modeling approach founded on high-level programming languages such as SystemC. It highlights the concept of separating communication from computation within a system.

In TLM notion, components are modeled as modules with a set of concurrent processes that calculate and represent their behavior. These modules exchange communication in the form of transactions through an abstract channel. TLM interfaces are implemented within channels to encapsulate communication protocols. To establish communication, a process simply needs to access these interfaces through module ports. Essentially, the interface is the very part separating communication from computation within a TLM system.

TLM defines a transaction as the data transfer (i.e. communication) or synchronization between two modules at an instant (i.e. SoC event) determined by the hardware/software system specification It could be any structure of word or bit, for example, half-word transfers between two peripheral registers or full image transfers between two memory buffers. The definition of transaction can be refined as a structure that is bus-protocol aware, i.e. it may include information as bus width or burst capability. Such refinement could be very helpful for SoC architects who perform fine analysis of arbitrations in SoC interconnection.

TLM proves itself a reliable methodology to wrestle with the clogging SoC bottlenecks. Throughout the SoC design cycle, it serves as the *unique* reference across different teams for three strategic activities:

- early software development;
- architecture analysis;
- functional verification.

In the perspective of durable progress, TLM leads SoC developers to a number of benefits towards productivity and time-to-market breakthrough. Not only work consistency is assured across different teams through the unique reference of TLM, modeling efforts are also vastly rationalized. Naturally, TLM will induce both cost- and time-efficient SoC project management in the long run. Last but not least, TLM indirectly encourages personnel interaction through cross-team communication. Our approach combines clock-less and yet bit-true and address-true, resulting in a single transaction level modeling that enables multi-disciplinary teams joint work for SoC hardware/software design and verification project.

3.1 Overview of the Novel Design Flow

The novel SoC design flow comprises two parts: standard RTL-to-layout flow *plus* system-to-RTL extension. Figure 1-2 presents this newborn flow with the position of TLM clearly indicated.

Referring to the same figure, a given SoC project generally starts from customer specification where system requirements are well identified. These preliminary requirements are then written as paper specification. Based on the specification, system architects perform hardware/software partitioning for configuring optimal system architecture.

TLM finds its place right after HW/SW partitioning. Once the TLM platform is completed, the flow enters the concurrent hardware/software engineering phase. In this phase, the TLM platform serves as the unique reference for software and architecture teams to conduct early software development and coarse-grain architecture analysis respectively. It also serves the verification team to develop the verification environment and its associated tests so as to verify the RTL platform once it becomes available.

Meanwhile, hardware designers develop RTL design of the system that produces a SoC RTL platform. As a result, a veritable hardware/software co-design is attained. The HW/SW co-design is one of the most remarkable differences between the novel and classic SoC design flow.

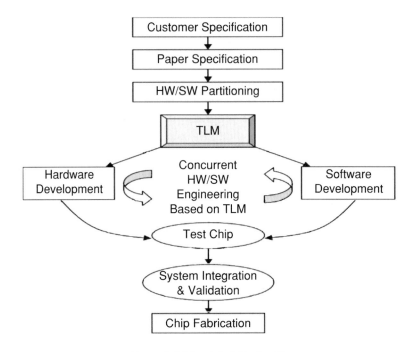

Figure 1-2. Novel SoC Design Flow

Once the RTL platform is available, various tasks could be conducted such as verifying its compliance with the intended performance, hardware verification, and low-level software integration with the hardware. These tasks perform concurrently with emulation setup, synthesis, and back-end implementation. The well-verified hardware design will then be taped-out for test chip fabrication. As the first test chip is ready, software design such as device drivers, firmware, or simplified applications will have also been verified with good level of confidence. Since both hardware and software designs are thoroughly verified, the novel SoC design flow will certainly increase the probability to achieve first-time silicon success.

3.2   Triple Abstraction

Our new design flow defines a structure of triple abstraction as follows:
1. SoC Functional View;

2. SoC Architecture View;
3. SoC Micro-architecture View.

The three views have complementary objectives to balance the need for both high simulation speed and accuracy. The triple abstraction can be integrated gracefully into the SoC design cycle without creating any conflict.

- **SoC Functional View**

Being the highest abstraction in the flow, SoC functional view abstracts the expected behavior of a given system in the way that users would perceive. It is an executable specification of the system function composed of algorithmic software. SoC functional view is developed without considering the implementation details at all, i.e. it contains neither architecture nor address mapping information. Performance figures are usually specified separately as paper specification.

- **SoC Architecture View**

Further down in the flow is SoC architecture view where TLM platform is conceived. This view captures all the necessary information to develop the associated software of a given SoC. Thus, hardware-dependent software can be developed and validated based on this abstract view long before it can be executed on a SoC physical prototype.

During the early design phase, this view also serves system architects as a useful means to obtain quantitative figures in determining optimal architecture that will best fit the customer requirements.

Another interesting point about SoC architecture view is its role of providing a reference model for verification engineers. Such reference is indeed the "golden model" for verification engineers to generate functional verification tests that will be applied on implementation models. These verification tests help to verify whether the system-under-design functions are in accordance with its expected behavior.

- **SoC Micro-architecture View**

The lowest level of the triple abstraction is SoC micro-architecture view. This abstract view captures all the required information to perform timed and cycle-accurate simulations. The prevalent modeling practice for this view is coding at register transfer level with hardware description language such as VHDL or Verilog. These models are very often made available since they are the most common input for logic synthesis to date.

SoC micro-architectural view is engaged in two key missions. First, it debugs and validates low-level embedded software in the real hardware simulation environment. The goal is to debug and integrate device drivers into the target operating system before the first test chip, or even before the hardware emulator is accessible. Second, this view helps greatly in SoC micro-architecture validation. System embedded software is normally optimized with the hardware being configured accordingly in order to sustain real-time requirements of an application. In case of insufficient performance, SoC architecture could be upgraded to match these requirements by using RTL views for any part requiring cycle accuracy.

[1] gives a good illustration of the activities based on SoC micro-architecture view and [12] describes a way to use multi-level models in a refinement flow.

3.3   Advanced Usage of TLM

SystemC TLM has so far been deployed for functional testing of embedded software and hardware RTL, as well as for hardware architecture studies. Certain algorithm teams use TLM APIs to structure their algorithms developments and make them readily usable by other teams of the SOC project such as functional verification engineers.

Architects will soon benefit from several advances in TLM needed for their work. Some of these advanced features enable further TLM deployment in software development teams. In addition, they contribute to drastic improvements in the way that the SoC integrator (semiconductor company) and its customer (system company) can cooperate for efficient definition of next generation SoCs.

A first step is the automated assembly of TLM, RTL, and mixed top-level SoC netlists from libraries of IP views described in SPIRIT XML standard format. This minimizes significantly the effort in assembling SoC simulations for architects, designers, functional verification engineers, and embedded software developers.

Using TLM not only for functional simulation but also for estimations of SoC performance, is progressing based on advances in adequate structuring of additional time-oriented wrappers around existing TLM models: the PVT models [7]. PVT modeling enables the architect to perform initial estimates right in the beginning of the project, without RTL or cycle-accurate IP models, even with rough functionality or algorithms coming from earlier studies. The precision of estimates can be increased anytime along the SoC project according to the needs and the updates from on-going hardware and software design. Early estimations of power consumption enabled by real software running on TLM model of SoC before RTL is available to further assist the architect. TLM thus facilitates the development of power-aware SoC software ahead of the hardware. Moreover, using TLM with place and route tools early in the SoC project could help in closing back-end tasks in a timely manner. Alternatively or in a complement way to address the P&R issues of large new SoCs, the Globally Asynchronous Locally Synchronous

(GALS) architecture is a natural fit for TLM modeling in all design and verification steps as demonstrated in a real taped-out GALS SoC [13].

Another key improvement is the ability to perform multi-tasking or multi-threaded embedded software architecture analysis on the envisaged SoC (single or multiple-processor). SystemC 2.0 has a non-preemptive scheduler that forbids a direct use of SystemC threads to model software tasks. This scheduler has other interesting properties for hardware design such as repeatability. One could however use a processor ISS linked to the SystemC simulation to run multi-tasking embedded software; but the ISS speed would very likely prevent any large pieces of software or data-intensive software (e.g. video processing) from running. Another approach is the socket connection between the SystemC simulator and tasks of the embedded software running natively compiled for the workstation and executing without ISS. Nevertheless, the socket connection and the workstation process or thread switching limit the speed. On top of it, the embedded software must be written according to certain guidelines.

There are two solutions. First, modify SystemC kernel, which may not be currently suitable to run existing hardware models with a range of commercial simulators that support existing OSCI SystemC semantics. Second, develop a scheme with special C++ wrapper enabling native compilation of unchanged multi-tasking embedded software (typically C) and fast execution in a linked SystemC 2.1 standard simulation. Advances in the last option are promising: multi-tasking software of several hundred thousands of C source code lines can be ported on TLM SoC simulation in one and a half day executing in the mega-Hertz range of simulation speed.

Regarding the equivalence of functionality between TLM and RTL models of SoCs, advances are being made in the areas of automatic comparison of TLM and RTL simulations despite their drastic difference in abstraction levels. The formal proof of TLM models is an on-going research topic that provides encouraging initial results.

The next area is the specification-to-TLM flow for hardware/software co-design, before and after hardware/software partitioning. Before partitioning, the OSCI TLM standard could be used to create a point-to-point, address-less functional yet concurrent SystemC model, reusing IP behaviors of the C code from application algorithm engineers. Tools should use this model in conjunction with hardware/software partitioning hypothesis, along with IP interfaces such as registers and the sub-system address-map information formalized in standard SPIRIT XML format [14]. The latter automatically wraps the C behavior in the address-mapped TLM model of hardware as described in earlier section, which is useful for running the embedded software by software and functional verification engineers.

Such model transformation may benefit from Model Driven Architecture (MDA) techniques, exploiting more formal specification capture than usual text thanks to notations like UML completed with suitable semantics extensions. A formalized initial specification above TLM will also benefit the formal verification of the SoC hardware and software design all the way from the refinement or generation design flow down to TLM, then RTL and constraint-driven generation of real-time embedded software.

In addition, new SoC architectures will provide further opportunities to profit from TLM models capabilities. Given the variety and complexity of fast evolving applications requirements along with sky-rocketing costs of design, verification, and mask production of nanometer SoCs, trends for new high-end SoCs are having more functionalities in a number of processors with the embedded software rather than dedicated hardware. This is also due to costs of programmable logic for raw processing, network-on-chip configuration, and coprocessor extensions.

Some intelligent load-balancing scheme will also be required [8, 15]. The optimal SoC hardware/software architecture for a given range of applications (i.e. an application domain) cannot be studied for sure using traditional combination of spreadsheet, some existing RTL, and ad-hoc partial models in C++. A complete modeling and simulation scheme with relevant analysis tools is needed, which is the exactly the sweet spot of SystemC TLM. The computing power will certainly exceed the one available on a single workstation. Thus, we have worked for parallelizing the SystemC simulation kernel to run large models of networked SoCs comprising multiple processors and complex hardware blocks. Such simulations can run on symmetric multiprocessor servers (SMP). Depending on the mapping efficiency of the SoC functionalities on the simulation computers, it could also run on a clustered Non-Uniform Memory Architecture (NUMA) configuration for supercomputing. Save and restore features in upcoming SystemC simulators will also help next-generation large-scale SoCs to simulate in an acceptable duration, e.g. software developers can debug a specific corner case that happens after million equivalent cycles.

The final breakthrough that will finish establishing TLM as the entry point of the SoC design flow is obviously automatic generation of synthesizable RTL from TLM models. This is a visible progress seen in next commercial tools offers. The OSCI standard of SystemC TLM modeling and the OSCI SystemC RTL synthesizable subset specification are also contributing to make this happen. Formalized initial specifications, e.g. in UML complemented with suitable semantics, down to TLM then RTL will be needed to reach the ultimate goal of affordable automated formal design and verification of SoCs. TLM acts as the intermediate pillar that reduces the

Specification-to-RTL gap into two smaller manageable gaps: Specification-to-TLM and TLM-to-RTL and embedded software.

One could envision a world full of networked, field-configurable heterogeneous multi-processor NOC-based SoCs with some FPGA areas suitably sized and located for a given range of applications. Such SoCs will be able to offer hardware and software functionalities downloaded from the Internet on demand by end-users, for instance, in multimedia mobile (communicating PDAs/games/video consoles) or embedded home and car equipments. A reliable performance service can still be assured after the download of additional new hardware/software applications in the device thanks to online architecture constraints TLM-based fast analysis (with automatically generated TLM/RTL and TLM/software adaptors), and downloadable configurations optimized for user-selectable trade-offs of performance, security, and power consumption.

REFERENCES

[1]   C. Chevallaz, N. Mareau, and A. Gonier, "Advanced Methods for SoC Concurrent Engineering," in Proc. of Design, Automation and Test in Europe Conference (DATE'02), 2002, pp. 59-63.

[2]   G. Martin, H. Chang, L. Cooke, M. Hunt, A. McNelly, and L. Todd, "Surviving the SoC Revolution – A guide to Platform-Based Design," Chapter 4, Kluwer Academic Publishers, 1999.

[3]   P. Magarshack and P. G. Paulin, "System-on-Chip Beyond the Nanometer Wall", in Proc. of 40th Design Automation Conference (DAC), Anaheim, June 2003.

[4]   D. Gajski, J. Zhu, R. Domer, A. Gerstlauer, and S. Zhao, SpecC: Specification Language and Methodology, Kluwer Academic Publishers, 2000

[5]   M. Baleani, A. Ferrari, A. Sangiovanni-Vincentelli, and C.Turchetti. "HW/SW Codesign of an Engine Management System", in Proc. of Design, Automation and Test in Europe Conference (DATE'00), 2000.

[6]   A. Clouard, G. Mastrorocco, F. Carbognani, A. Perrin, and F. Ghenassia, "Towards Bridging the Precision Gap Between SoC Transactional and Cycle-Accurate," in Proc. of Design, Automation and Test in Europe Conference (DATE'02), 2002.

[7]   OSCI Standard for SystemC TLM, Available at HTTP: http://www.systemc.org

[8]   P. G. Paulin, C. Pilkington, M. Langevin, E. Bensoudane, and G. Nicolescu, "Parallel Programming Models for a Multi-Processor SoC Platform Applied to High-Speed Traffic Management," in Proc. of International Conference on Hardware/Software Codesign and System Synthesis (CODES+ISSS), 2004 (Best Paper Award).

[9] A. Haverinen, M. Leclercq, N. Weyrich, and D. Wingard, "SystemC-based SoC Communication Modeling for the OCP Protocol," [Online document] 2002 Oct 14 (V 1.0), [cited 2004 Nov 5], Available at HTTP: http://www.ocpip.org/socket/whitepapers/

[10] SCE-MI Standard for Transaction-based Co-emulation, Information available at ACCELERA website: http://www.eda.org/itc

[11] A. Clouard, K. Jain, F. Ghenassia, L. Maillet-Contoz, and J.P. Strassen, "Using Transactional Level Models in a SoC Design Flow," Chapter 2, SystemC Methodologies and Applications, Ed. W. Müller, W. Rosentiel, J. Ruf, Kluwer Academic Publishers, 2003, pp. 29-63.

[12] G. Nicolescu, S. Yoo, and A. Jerraya, "Mixed-Level Co-simulation for Fine Gradual Refinement of Communication in SoC Design," in Proc. of Design, Automation and Test in Europe Conference (DATE), 2001, pp. 754-759.

[13] E. Beigné, F. Clermidy, P.Vivet, A. Clouard, and M. Renaudin, "An Asynchronous NOC Architecture Providing Low Latency Service and its Multi-level Design Framework," in Proc. of ASYNC, 2005.

[14] SPIRIT Consortium Website at http://www.spiritconsortium.org

[15] P. G. Paulin, C. Pilkington, and E. Bensoudane, "StepNP: A System-Level Exploration Platform for Network Processors," IEEE Design & Test of Computers, vol. 19, no. 6, Nov.-Dec. 2002, pp. 17-26.

Chapter 2

TRANSACTION LEVEL MODELING
An Abstraction Beyond RTL

Laurent Maillet-Contoz and Frank Ghenassia
STMicroelectronics, France

Abstract: Transaction level modeling (TLM) is put forward as a promising solution above Register Transfer Level (RTL) in the SoC design flow. This chapter formalizes TLM abstractions to offer untimed and timed models to tackle SoC design activities ranging from early software development to architecture analysis and functional verification. The most rewarding benefit of TLM is the veritable hardware/software co-design founded on a unique reference, culminating in reduced time-to-market and comprehensive cross-team design methodology.

Key words: transaction; untimed model; timed model; initiator; target; channel; port; concurrent processes; timing accuracy; data granularity; model of computation; system synchronization; functional delay; annotated model; standalone timed model.

1. THE REVOLUTION

1.1 Call for Raising Abstraction Level

Squeezed by the ever-increasing SoC design complexity, cost, and time-to-market stress, the much-perturbed SoC industry is longing for a solution. The key to this solution is to improve the design productivity through a more reliable design methodology within a shorter design time-frame.

Forwarding critical software development earlier in the SoC design flow is unquestionably helpful to reduce the design cycle time. Such advance implies indeed a hardware/software co-design wherein the software is developed in parallel with the hardware for earlier system integration.

To cope with the rising SoC complexity, a much more rigorous methodology is sought after to assure the reliability of SoC performance at

F. Ghenassia (ed.), Transaction Level Modeling with SystemC, 23-55.
© 2005 *Springer. Printed in the Netherlands.*

an earlier stage of the design cycle. A favorable approach is the architecture exploration that analyzes the potential effect of the realistic traffic performed by a system.

Pulling all these factors together, raising the level of abstraction above RTL in the overall SoC design and verification flow has appeared to be a promising solution for the SoC industry.

1.2 Attempts at Raising Abstraction Level

Bear in mind that any attempt made to raise the abstraction level is always a game of balancing the trade-off between the speed and accuracy of a potential simulation model. Our development effort has of course witnessed this game from tip to toe. Before tackling the subject of abstraction level, it is worth considering what the two extreme ends of the SoC design flow could offer.

First, consider the algorithmic model at the highest end of the flow. A complex design usually begins with the development of such a functional model. As an example, a digital signal processing oriented design will have a dataflow simulation engine as its algorithmic model. Since it only captures the algorithm regardless of the implementation details, an algorithmic model has a huge advantage in its high simulation speed. In spite of this, an algorithmic model has no notion of hardware or software component; it models neither registers nor system synchronizations related to SoC architecture. This model therefore cannot fulfill the need of executing the embedded software.

On the other end of the design flow, a pure logic simulation can take place at the register transfer level (RTL). In a conventional SoC logic simulation, RTL models written in hardware description language (HDL) such as VHDL and Verilog are employed as the system hardware. If a processor model is necessary, a design sign-off model (DSM) will typically be used. The advantage of the logic simulation is evidently its great fidelity to the real implementation, i.e. accurate SoC functional and performance analysis. This is nonetheless a price too expensive to pay in terms of the lengthy simulation time. The time consumption has actually further worsened lately due to the high SoC complexity that requires a longer RTL development phase. Moreover, a pure logic simulation cannot execute any software in a reasonable amount of time. A system can only integrate its associated software for observation and analysis rather late in the design flow. Since the breadboard is usually almost ready at this point, any system modification will certainly be too costly at this stage.

In brief, an in-between solution has to be resolved for which three fundamental criteria must always be respected as the doorway to early software development and architecture exploration:

1. *Speed*. The potential model must simulate millions of cycles within a reasonable time length. The target activities frequently involve a very large scale of simulation cycles. Some of them may entail user interactions that could probably slow down the process. It is unacceptable and unaffordable to wait for even just a day to complete a simulation run.

2. *Accuracy*. Although speed is an interesting advantage to enhance, the potential model should sustain a certain degree of accuracy to deliver reliable simulation results. Some of the analyses may require full-cycle accuracy to obtain adequate outcomes. As a rule of thumb, the potential model should at least be detailed enough to run the related embedded software.

3. *Lightweight Modeling*. Any other modeling effort in addition to the compulsory RTL modeling for hardware synthesis must be kept insubstantial to optimize the overall SoC project cost. The potential model should be, for this reason, a quick-to-develop model at a considerably low effort.

Collected here are some attempts to raising the abstraction level. Brief descriptions are provided for these attempts, including hardware/software co-verification, cycle-accurate model, and temporal model.

- **Hardware/Software Co-Verification**

The concept of hardware/software co-verification is suggested for reducing the critical SoC design time and cost to overcome the limitation of pure logic simulations. The underlying idea of this concept aims at leading hardware/software integration, verification, and debugging to an early phase of the design cycle before the real hardware is available.

RTL models remain the hardware models in a co-verification platform. An obvious difference from pure logic simulation is that co-verification uses a faster processor model, i.e. Instruction Set Simulator (ISS). This is an instruction-accurate model developed in C language at a higher level of abstraction.

The co-existence of hardware and software during the SoC verification process is the essence of co-verification. While the hardware platform is connected to a logic simulator, a symbolic debugger links the associated software program to the ISS for its execution on the platform. Such co-operation offers a simultaneous controllability and visibility over both hardware and software to analyze the system behavior or performance. The simulation speed is of orders of magnitude higher than the one of logic simulation. Since the breadboard is not manufactured yet, any modification of the system hardware or software at this stage will be both time and cost-efficient.

Despite the numerous benefits yielded by the co-verification, it is still too long to wait for the development of RTL hardware models before the co-verification can be conducted. The time pressure has pushed us to tackle another approach: cycle-accurate model.

- **Cycle-Accurate Model**

  This attempt tries to replace the non-processor hardware parts by a model residing at higher level of abstraction. The prospective model could be developed using high-level programming languages such as C. Compared to RTL models, this model is less precise. It is sensitive to whatever happens at the interval of each clock cycle, which is more than enough for software verification but not providing any synthesizable description.

  With the emerging C-based dialects that support hardware concepts, it seems convincing that cycle-accurate models developed in a C-based environment could meet the three criteria mentioned earlier for raising the abstraction level. However, this hypothesis has stumbled upon a few obstacles [1-4]:

  a) Most of the information captured by cycle-accurate models is unavailable in IP documentation but only in the designer's very mind and the RTL source code itself! Consequently, RTL designers have to invest much time to keep modeling engineers informed; otherwise modeling engineers must reverse-engineer the related RTL code. Either way ends up being a tedious and time-consuming process without actually solving the issue.

  b) Cycle-accurate models can simulate merely an order of magnitude faster than the equivalent RTL models, which is really just too close to the speed of VHDL/Verilog models.

  Not only is simulation speed too slow to run a significant amount of embedded software in a given time-frame, the development cost is also too dear to compensate for the negligible benefits of cycle-accurate models. In addition, architects and software engineers do not require cycle-accuracy for all of their activities; for instance, the software development may not involve any cycle-accuracy until engineers work on the optimization.

- **Temporal Model**

  Instead of balancing speed and accuracy, the temporal model is attempted as quite a different approach to raise the abstraction level. This model is mainly opted for the performance analysis of a system. While timing analysis is the focus of temporal models, analytical accuracy is forgone.

Some efforts were given in the development of the temporal model. The resulted model provided extremely high simulation speed but with little or virtually no functional accuracy guaranteed. The temporal model is thus far from being the ideal solution to our need of raising the abstraction level.

1.3   Birth of Transaction Level Modeling

Through our different attempts for raising the abstraction level, we have concluded that the most compelling resolution is to adopt the famous "divide and conquer" approach. This approach counts on two complementary environments as the best bid to balance the trade-off between simulation speed and accuracy, i.e. transaction level modeling (TLM) platform and register transfer level (RTL) platform.

- **SoC TLM Platform**
  TLM platform is intended for early SoC exploration in the design flow at a relatively lightweight development effort. It is a transaction-based abstraction level residing between the bit-true cycle-accurate model and the untimed algorithmic model. Our development work has demonstrated that SoC TLM platform makes an excellent complement to RTL platform as an adequate trade-off between simulation speed and accuracy. On top of the untimed functional TLM, it is also possible to add timing annotations to TLM platforms for early performance analysis without paying the cost of cycle accurate models.

- **SoC RTL Platform**
  RTL platform aims for fine-grain SoC simulations at the expense of slower simulation speed and later availability. It applies cycle-accurate HDL models for a detailed timing analysis.

The idea of "divide and conquer" proves itself an extremely efficient modeling strategy. With the high modeling and simulation speed offered by TLM platforms, potential users could quickly accomplish a systematic analysis for a given SoC as the first approach. A comprehensive timing analysis based on RTL platforms will follow afterward to provide results that are more accurate. Hence, this complementary characteristic enables a system-under-design to go through rapid methodical study as well as in-depth exploration. Figure 2-1 gives the efficiency levels of the different modeling strategies, including RTL, cycle-accurate model (CA), and TLM. It shows clearly how TLM helps the concept of "divide and conquer" become a success through its high modeling and simulation speed.

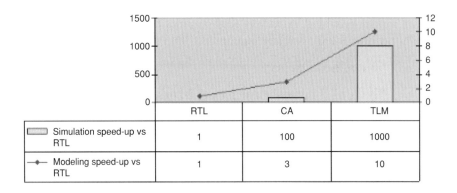

Figure 2-1. Efficiency of Modeling Strategies

A question wondering in your mind now could probably be "Why would TLM be so interesting compared to other rival propositions?" The answer is that we have successfully identified the appropriate level of abstraction, *TLM*, which has a description usable for embedded software development and early architecture analysis thanks to its adequate trade-off between simulation speed and accuracy.

Most of the propositions available in the field use proprietary C-based languages such as SpecC, Hpascal or HardwareC to implement cycle accurate models. High-level models, on the other hand, are either expensive solutions sold by CAD vendors or limited versions reserved for academic applications. Although these high-level models give temporal view of a system, they are not precise enough to develop any embedded software.

Before considering the advantages that TLM has to offer, its very distinct point from other propositions is the use of SystemC *-an open-source programming language-* that suggests a free of charge development environment for a tangible solution.

SystemC provides a foundation to model hardware and software of a system based on a single language. It is an object-oriented approach built on top of C++ as a set of classes. A system conceived by SystemC demonstrates particular characteristics in concurrency, reactivity, distributiveness, timing, and data types. Further details of TLM modeling techniques using System C will be discussed in Chapter 3.

The remainder of this chapter presents a zoom-in discussion on TLM ranging from its principles to its battle against the SoC design bottlenecks.

2. PRINCIPLES OF TLM

2.1 Terminology

TLM offers a new SoC design methodology at a higher abstraction level above RTL, i.e. a transaction-level modeling technique intended for digital electronic systems.

In a digital electronic system, every single component is composed of a finite set of states and a series of concurrent behavior. TLM models each of these components as a ***module***. The internal states of a component are represented by a set of variables defined within the scope of the corresponding TLM module, whereas the different behavior pieces of the component are modeled by a collection of ***concurrent processes*** or ***threads***, which can be executed in parallel.

Just like the components of a SoC, TLM modules are gathered to form a TLM system. Through a specific TLM communication structure, namely ***channel*** or ***interconnect***, communications are established between modules. Depending on the accuracy level required by the corresponding simulation, a channel could be a simple router, an abstract bus model, a network-on-chip model, or some other structures. This is essentially the very part that separates communication from computation in TLM modeling.

Modules and channels are bound to each other by means of communication ***ports***. Once they are bound together, data can be exchanged between them to perform the expected system behavior. Potentially, data can also be communicated between modules and test-benches.

The term ***transaction*** denotes the set of data being exchanged. A ***master*** or ***initiator*** is a module that initiates transactions in a system, while a ***slave*** or ***target*** is a module that receives and serves transactional requests. Any consecutive transactions may have various sizes of data transfer. This variable size corresponds to the amount of data being exchanged between two occurrences of system synchronization.

System synchronization is an explicit action between at least two modules (potentially test-benches) that need to coordinate or manage some behavior distributed over them. Such co-operation of different modules is vital to assure the predictable system behavior.

Since it is the only mechanism available for synchronizing the different processes in a system, the explicit system synchronization is compulsory to ensure a proper deterministic SoC TLM behavior. An example of system synchronization is the interrupt raised by a direct memory access (DMA) to notify a transfer completion within a system.

2.2   Modeling Approach

The terms of TLM defined in the last section can be attained through an appropriate electronic system level (ESL) modeling approach. The right candidate to do this job is a high-level programming language that is capable of developing not only a plain software program, but also of modeling electronic hardware at the conceptual level without describing the real implementation. The potential candidates include SystemC, SpecC, Hpascal, System Verilog, HardwareC, and the like. In our opinion, SystemC is the best candidate and we therefore rely on it for all of our TLM models.

As discussed earlier, a SoC component is modeled as a module in TLM. The primary modeling effort lies in the internal computation of the given hardware block at the functional or behavioral level. The input and output of the block as well as its synchronization are to be modeled. None of the micro-architectural implementation details should be included, i.e. neither internal pipelines nor structures are modeled. To sum up, TLM modules representing SoC hardware blocks or IPs must hold the three characteristics stated below:

1. bit-true behavior of the component;
2. register-accurate interface of the component;
3. system synchronizations managed by the component.

A complete SoC TLM platform is constructed by instantiating and binding different modules and channels together. Once the platform is integrated, SoC simulation is performed by executing the related embedded software either as native or cross compilation. The earlier is executed on a simulation workstation for fast simulation speed, while the latter is executed on the embedded processor architecture, i.e. ISS, for precise simulation accuracy.

To ensure a proper system functional behavior in TLM SoC simulation, there are two essential points that deserve attention in the modeling process. First, all the data transactions must be blocking i.e. the thread that initiates the transaction will resume its execution only if the current transaction is completed. Second, all the occurrences of the system synchronization must be potential re-scheduling points in a simulation environment in order to guarantee an accurate simulation of the concurrency. The system synchronization could be modeled by specific means such as event, signal, and interrupt; or by data-exchanges such as polling. If any of these potential system synchronizations causes a call to the simulation kernel, it enables the scheduler to activate other modules. Hence, the simulated system will behave correctly in line with its functional concurrency.

The essence of working out an appropriate model at transactional level lies in the good sense of deciding where and when to implement system synchronization. If too many synchronized points are inserted, the model will tend to be too close to cycle-accurate or RTL models that will not help to gain much simulation speed. Contrarily, if too few synchronized points are implemented, the model may run the risk of having incorrect system execution.

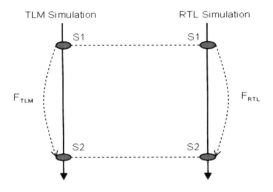

Figure 2-2. TLM vs RTL Simulation

Consider the two simulations depicted in Figure 2-2, which are correspondingly the RTL and TLM simulations for a given system. The evolution of the system from the first stable system state, S1, to the next stable system state, S2, is represented by F_{RTL} and F_{TLM} respectively. Indeed, S1 and S2 are two partial observation points in simulation, i.e. two synchronization points.

F_{RTL} is a collection of all necessary cycle-accurate computations to bring S1 to S2. These calculations are implemented by a set of clocked processes that represent the system micro-architecture. Upon each clock cycle, these processes are activated in the simulation kernel for execution; and that will consequently involve countless of context switches.

On the other hand, F_{TLM} is an equivalent function to bring S1 to S2 but without any clock implementation. Computations are defined by some high-level programming languages such as C or C++. There is principally sequential execution of programming codes between S1 and S2. Compared to RTL simulation, it involves much fewer parallel executions of processes. As a result, there are relatively less context switches involved.

Recall the efficiency levels of different modeling techniques illustrated in Figure 2-1, the simulation speed-up achieved by TLM is vastly ahead of RTL up to a factor of 1000. Indeed, this speed-up correlates directly with the number of processes and context switches activated between two occurrences of system synchronization by RTL simulation *but not* by TLM simulation kernel.

2.3   Modeling Accuracy

The modeling accuracy of a given modeling approach indicates the precision or correctness of the model in replicating the intended behavior and activities of a system-under-design. For any modeling strategy in the SoC design flow, there are two decisive factors to determine the degree of modeling accuracy:

1.   *Granularity of Communication Data.*
     This criterion reflects the fineness or coarseness of the data carried by the communication structure of a model. The data granularity can generally be categorized into three levels, i.e. application packet, bus packet, and bus size, in the order of increasing accurateness. The transfer of a video IP helps to illustrate the idea of data granularity. If the IP has a frame-based algorithm, a coarse granularity at application packet could be modeled as a frame-by-frame transfer. A finer granularity at bus packet level can be represented by a line- or column-based transfer, or a macro-block transfer consisting both lines and columns. The finest grain at bus size level will be the pixel-based transfer of the video.

2.   *Timing Accuracy.*
     Timing accuracy determines the fidelity of a model to the intended timing behavior. It can be conceptually perceived as a scale of two extremes, i.e. untimed level and cycle-accurate level. Moving from the untimed end towards the cycle-accurate end will increase the timing accuracy of a model. Any level falling in between the two ends is considered as *approximately* timed level.

Just as any other modeling strategies in the SoC design flow do, the TLM approach naturally revolves around the two factors above to decide its modeling accuracy. Guided by these criteria, we have conceived two fundamental classes of TLM to date through our development effort:

- Untimed TLM.
- Timed TLM.

The untimed and timed TLM are models tailored for distinct purposes. The ultimate goal is to create a unique platform that simulates two different models according to user needs.

The untimed TLM is an *architectural* model targeted specifically at early functional software development and functional verification where timing annotations are not compulsory conditions. The high simulation speed is the objective of this model. Since the untimed TLM serves primarily programmers, it is hence given another name as *programmer's view* (PV).

On the other hand, the timed TLM is a *micro-architectural* model containing essential time annotations for behavioral and communication specifications. It is relatively a less abstract model located lower in the SoC design flow. The focus of timed TLM is the simulation accuracy required by real-time embedded software development and architecture analysis. Hence, the timed TLM is also known as *programmer's view plus timing* (PVT).

Figure 2-3 gives a glimpse at the modeling accuracy of the untimed and timed TLM with respect to other conventional models in the SoC design flow, including register transfer level (RTL), bus cycle accurate (BCA), and cycle accurate (CA) models.

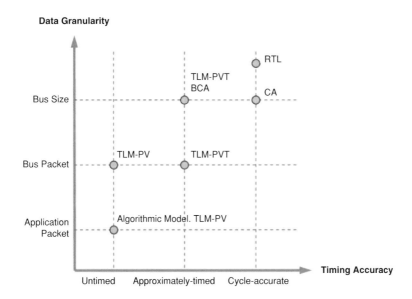

Figure 2-3. Modeling Accuracy of Various Approaches

3.   UNTIMED TLM

3.1   Introduction

The untimed TLM is a level particularly conceived for serving software programmers and verification engineers in early functional software development and functional verification. Timing annotations are insignificant at this untimed level; thus, none of the information related to the micro-architecture of the component or IP-under-design should be included.

For the same reason, any information related to the interconnect topology and arbitration law will not be captured in the untimed TLM. The internal states of a component are modeled by using appropriate internal variables.

Certain information, for instance, the register bank or memory content of a given component, is made available and accessible to the outside world through a well-defined Application Programming Interface (API). The communication API is a blocking API that provides a particular interface to supervise full data transfer.

The granularity of the data transferred should correspond to the modeling level related to the target application. For example, data transfer of an image-processing block should be modeled at the frame level, i.e. one frame being transferred at a time rather than creating transfers of the bus width.

3.2   Model of Computation

The untimed TLM has absolutely no timing information related to the micro-architecture, i.e. there is *no* clock in an untimed TLM system. Since it has no clocked timing regulation, all processes are executed concurrently to access any of the system resources at the same time instant. Yet, the system must demonstrate a correct behavior during the parallel execution of concurrent processes. This implies that untimed TLM systems must respect a certain degree of process execution order to guarantee a proper system functional performance.

To fulfill this requirement, the untimed TLM employs a specific model of computation with the following characteristics:
1. concurrent execution of independent processes;
2. respect for causal dependencies between processes using system synchronization;
3. bit-true behavior;
4. bit-true communication.

3.2.1   System Synchronization

A system must clearly characterize the causal relation between its different processes in order to assure deterministic system behavior. The explicit system synchronization is therefore implemented within a system to respect such causal dependencies. The system synchronization only defines a partial execution order for SoC internal events, i.e. a partial execution order between the different processes in the whole system. In other words, any particular execution order among all of the processes is permitted as long as their causal dependencies are well respected.

To better illustrate this idea, consider three processes in a given system, P1, P2, and P3, as depicted in Figure 2-4. Assume that each process denotes a thread for a particular module in the system.

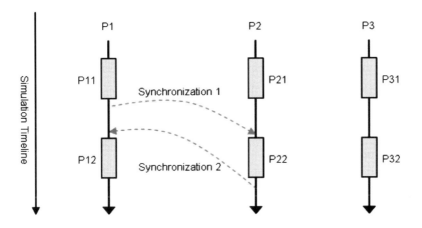

Figure 2-4. System Synchronization between Processes

The full execution order within each of these processes is represented by their own internal synchronized events:

  a) P11 → P12 for process P1
  b) P21 → P22 for process P2
  c) P31 → P32 for process P3

Bear in mind that this "full" order is only a locally complete order within each process. It is indeed a "partial" execution order from the point view of the overall system execution. Besides, there are two occurrences of system synchronization between P1 and P2, which give additional constraints to the overall system execution order:

  d) P11 → P22
  e) P22 → P12

The constraints of execution order stated from (a) to (e) clearly describe the causal dependencies that must be respected within the system. The three processes can be executed with any particular order as long as these causal dependencies are followed. Here are some examples of the different *overall* system execution order (which are also known as process interleaves):

f) P21 → P11 → P22 → P12 → P31 → P32

g) P31 → P32 → P21 → P11 → P22 → P12

h) P11 → P21 → P22 → P31 → P32 → P12

The system synchronization is a mechanism to inform others or to get informed by others about some system state changes when these changes potentially influence the execution of some other parts of the system. In real hardware circuits, system synchronizations are modeled by means of interrupt signals, polling or mailbox. The TLM simulation will implement all of the system synchronizations as interrupts, mailbox or polling in line with the model of computation stated earlier. An abstract implementation of the various synchronization mechanisms, however, could be provided to better match with the considered level of abstraction.

According to its nature of informing or being informed, there are two kinds of synchronization. First, "emit-synchronization". This occurs when a process sends out a synchronization that may influence the behavior or state of other processes. Second, "receive-synchronization". This is a point where a process waits for an incoming event from the system that may influence its behavior or state.

Picture this: every synchronization point is a traffic light in a given system. Each of these "traffic lights" is associated to a certain condition; for instance, the occurrence of an event or the computation of a particular value. Once this condition is fulfilled, the green light will be on to allow the system to proceed to the next execution point. Otherwise, the red light is there to stop it. All these little "traffic lights" scattered in the system has a big mission: work hand-in-hand to guarantee a proper predicted system behavior.

An important employment of the system synchronization is the assurance of memory or data consistency. Here, the system synchronization prevents concurrent processes from reading data content at unknown state; it also prevents them from writing data at temporarily inaccessible memory area.

A direct beneficial impact of the system synchronization is the capability of executing any legal interleaves of processes without breaking the overall system synchronization. The system synchronization also serves as an efficient method to improve the validation of the system simulation model by allowing more process interleaves to be tested. The model of computation only requires the causal dependencies to be respected by the simulation. Thus, it is possible to randomize the process selection as long as the system synchronization does not define a full order of process execution. This is particularly useful in the case where simulation kernels do not provide random process execution.

All of the system synchronization points in a system must be explicitly modeled for a correct system behavior. If an untimed TLM system ever generates any simulation deadlocks or failures, the explanation will be the system synchronization not being explicitly modeled to the fullest, or simply badly designed. By slight chances, a system with incomplete synchronization modeling may appear to function as normal at certain values of clock frequency. It will however fail to perform at other clock frequencies. Undoubtedly, such incomplete modeling will adversely jeopardize a safe chip execution.

3.2.2   Process Execution

The concurrent execution of independent processes is one of the major characteristics of the untimed TLM. Simulation kernels are usually implemented in such a way that they offer repeatable process executions to simplify debug activities. Note that simulation kernels cannot give a deterministic execution of concurrent processes (even the language reference manuals cannot guarantee a deterministic execution of concurrent processes). It means that we cannot predict which process that the simulation kernel is going to start executing; but once the simulation is executed, the kernel will repeat the same execution order.

Although the repetitive feature of simulation can facilitate the debugging procedure, a single system execution order may not provide satisfactory validation coverage. In our last example of system synchronization, the overall system execution can start with any of the three processes. If the simulation only covers a single execution order, we would probably miss catching the bugs hidden in other execution orders! As an example, imagine another synchronization that imposes a constraint of executing P21 before P11. If the repetitive simulation kernel picks the system execution order of (f) or (g), the simulation will pass without detecting any error. An error, however, would have occurred in the system performance by following the execution order of (h) where P21 is not executed before P11.

To tackle this limitation, we must make sure that any execution order will conform to the system functional specification. An appropriate solution to increase the coverage of system execution orders will be extending the standard simulation kernel with a random function that shuffles all of the legal process interleaves. With such mechanism, it is feasible to verify all of the possible micro-architectures of a given architecture specification.

This definition actually corresponds to the implementation of asynchronous processes that use synchronization points to ensure a correct execution of the system. If one expects to cover all of the possible process interleaves as in the real-life system, it will obviously produce a huge number of combinations with lots of them being meaningless. Hence, it is worth-noticed that it is possible to reduce the indeterminism of concurrent process execution by introducing successive constraints in the untimed models based on their partial system execution orders.

A typical example is the integration of timing constraints that make sense at the functional level. The objective is to reduce the number of potential process interleaves by adding constraints in the selection of the various processes for the simulation. Here, the timing information is only related to functional constraints (e.g. a video application imposes to decode 30 frames per second), but no information on the micro-architecture is incorporated yet. The result is a decreased indeterminism, which reduces the simulation variants to be considered for the system validation. This will be further discussed in Section 4, *Timed TLM*.

3.2.3   Time-Independent Deterministic Behavior

This section explains how the computational model of the untimed TLM handles the constraints of process execution order without implementing timing characteristics.

Consider a fixed set of input stimuli for a given SoC. The system synchronization points implemented among the different processes will induce a deterministic behavior that is independent of any timing behavior during the simulation. Each of these processes follows a particular sequence as described in Table 2-1.

Table 2-1. Untimed Process Sequence

Step	Action
1	Activate or resume a process.
2	Read input data for control flow and data processing.
3	Computation.
4	Write output data if there is any of them.
5	Return to step 2 if more computation is required.
6	Synchronization:
	(a) if it is "emit-synchronization", return to step 2;
	(b) if it is "receive-synchronization", the process will be suspended.

When a process reaches step 6 in the untimed sequence, the component state will have already been fully defined, and the memory state modified by the process should be fully defined as well. Only when a process reaches step 6(b) of "receive-synchronization", it will be suspended. This is the only situation where a process needs an update of the system state that might influence its own behavior. As a result, the simulation kernel could by no means suspend a process by itself, i.e. the simulation kernel is not pre-emptive. This will definitely assure predictable process states and process controls, which are independent of any specific implementation of the simulation kernel.

Most of the time, a process could include many synchronization points and that will produce a very complex control flow graph with many possible activation-synchronization paths. Note that reducing the number of the descheduling points in a system model to the "receive-synchronization" can be very beneficial. While assuring a correct simulation of the SoC architecture, such reduction can greatly minimize the number of context switches compared to other computational models. Therefore, the kernel overhead is minimal, leading to the simulation speed close to the one of pure algorithm.

3.3 Modeling of Interrupts

Literally, interrupts mean disruptions that could result in certain consequences. For electronic systems, an interrupt is considered as a system event with side effects such as triggering a delayed management of processes or updating registers of interrupt-status.

Recall that system synchronization is very often implemented by an interrupt signal. In the untimed view, an interrupt is however an impulsive system event without any persistence. It is therefore inappropriate to model it using a signal. Instead, a dedicated TLM synchronization protocol with the following features is employed:

 a) immediate propagation of interrupts from an initiator to a target;
 b) notice of potential IP internal state change, i.e. status register update.

While developing untimed interrupt models, the first-in-first-out (FIFO) mechanism must not be implemented in the reception structure as it may cause serialization of concurrent events undesirable at that level. Upon the generation of an interrupt, the target IP may invoke a consequent effect out of its own scope. In that case, meticulous care must be taken so that another process *but* not the one generating the interrupt will handle the consequent effect. This will avoid changes in the system state caused by the process generating the interrupt in the Remote Procedure Call (RPC) coding style.

3.4   Insertion of Functional Delay

At the architectural level, it is still necessary to introduce some functional timing information, i.e. *functional delay*, when these delays are part of the system specification (e.g. a video decoder decodes 30 frames per second).

Sometimes, an untimed TLM IP is inserted with functional delay to implement implicit synchronization points related to specific timing information. As an example, a Liquid Crystal Display (LCD) controller with a screen-refresh frequency of every 1/30-second can be modeled without any explicit synchronization. It means that the untimed LCD controller can be created with implicit timings by adding some delay information and wait statements of specified time length into the model.

From the angle of computational model, such implicit timings bring additional constraints to the execution order of processes in the simulation, and thus reduce the set of possible process interleaves. As a result, the untimed model inserted with functional delay is created as an intermediate level between the purely untimed TLM and the timed TLM. Model developers should guarantee a flexible manipulation of this intermediate model by allowing users to easily enable or disable the annotated delay information. It must leave users enough room to switch back to a purely untimed model for validation purposes. Furthermore, this intermediate model should never cover any functional information related to the micro-architecture such as FIFO, Finite State Machine (FSM) related to cycle-accurate behavior, or any other implementation-dependent features.

Figure 2-5 illustrates the typical timelines of a process execution occurring in the untimed TLM. Two cases are demonstrated:

a) Simulation without functional delays based on a functional specification that only defines sequences of actions.

b) Simulation with functional delays based on a functional specification that defines some timing attributes such as UART baud rates.

Adding functional delays to an untimed model does not particularly influence the model of computation. Processes will still have activation, emit-synchronization, and receive-synchronization points. The execution order of various processes will be more constrained because the inserted functional delays restrict the set of potential process interleaves eligible for simulation. In other words, there are fewer choices of process interleaves for the simulation kernel at a given time instant.

Functional delays can suspend a process to induce the simulation kernel to choose other eligible processes for execution. This cause-and-effect phenomenon can influence the system state, but should *never* cause any system inconsistency from the perspective of computational model. The reason is that the system synchronization must *fully and explicitly* model all causal relations of a system. An error will otherwise arise in the system synchronization scheme, and that is considered as a serious bug in the SoC specification.

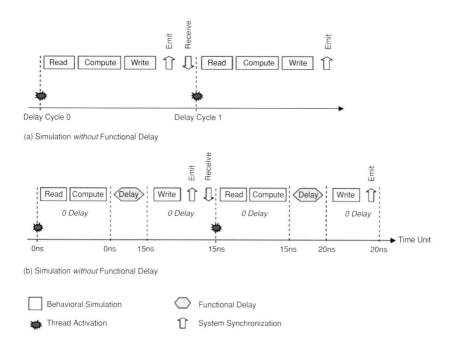

Figure 2-5. Simulation Timelines of Untimed TLM

Let us look into this statement more carefully through an example. Consider a system that is modeled by a group of processes denoted from P1 to Pn. Assume that a functional delay is inserted into the codes of P1, and that induces the simulation kernel to select another process, say P2, for execution. The system state could potentially be affected by the execution of P2. If that is the case, the global system state will have already changed when P1 resumes its execution.

Such global change of the system state should not influence the remaining execution of P1. This process should be able to continue its activities until it reaches the next functional delay or receive-synchronization point. If this interleaved execution of P1 and P2 happens to affect the remaining execution of P1, there is certainly a missing part of system synchronization somewhere between P1 and P2.

The adverse consequence of such incomplete modeling in the system synchronization is the dreadful inconsistent simulation result. This is the reason why the computational model of the untimed TLM obliges explicit modeling for every single system synchronization point in a given system.

Modeling engineers must insert functional delays into untimed models in such a way that the system synchronization can still manage to capture all the causal dependencies in a given system. This is a good modeling practice to assure the system stability, despite the variations of the clock frequency and the indeterminism of the micro-architecture (e.g. transaction latency on a bus depends on the bus load) in the sub-systems of a SoC.

3.5　Recommendation for Modeling Practices

Collected hereafter are our general recommendations for the untimed TLM modeling practices based on our experience in TLM development. Advices on implementation concerns are provided in Chapter 3.

1. Consider the intended uses of IPs on the final platform to efficiently determine how the corresponding TLM models should be written up.
2. To increase reusability, organize models in such a way that the algorithm can easily be updated, and reuse readily available standalone C models as much as possible. For the reason of code portability and management, these C models should never be replicated as "hardwired" copies in the TLM environment. Rather, they should be reused by means of wrappers or external function calls.
3. Determine the data granularity of models according to the algorithmic accuracy and the expected precision in terms of transfers. For example, the model of a video IP expecting frame-level input should be modeled with data granularity at frame level but not pixel level, despite the actual capability of the interconnect in the silicon. However, if there is a mismatch between the data granularity of the algorithm and the data layout in the memory according to the memory map, it will be the job of the TLM wrapper to generate the correct addresses so that the data is stored and retrieved from the correct memory locations.
4. Model all sorts of communication interfaces at bit-accurate level, particularly for register modeling.
5. Model all sorts of behavior at bit-accurate level.
6. Focus modeling with respect to the functional specification only, i.e. including *no* micro-architectural and clock-based information, resources, or details.
7. Model explicitly the system synchronization that affects the IP behavior.
8. Employ events within a model whenever that is appropriate for modeling the inter-process synchronization.

9. Utilize specific synchronization means such as synchronization protocols to model the inter-module synchronization.
10. Avoid implementing the process-activation based on a regular basis; the process-activation based on system activity is compulsory.
11. Ban uses of global variables.
12. Adopt good software implementation style to facilitate code debugging and maintenance, e.g. add comments in codes.

4. TIMED TLM

4.1 Objective

As far as we have discussed for the untimed TLM, the system synchronization only defines a partial order of the overall SoC internal events. The identification of the full order of SoC events is hampered by an indeterminism because the untimed TLM does not capture micro-architectural details, i.e. the timing behavior of the implementation.

The timing behavior of a component specifies the delay between each activation and synchronization-suspension. If this timing behavior is incorporated into TLM, the resulted timed model will be able to determine a full order of SoC events; hence leading to a complete specification of the implementation.

The main objectives for developing the timed TLM are:
- benchmarking of the performance of a given micro-architecture;
- fine tuning the micro-architecture;
- optimizing the software for a given micro-architecture to meet real time constraints.

Other objectives for implementing timed TLM models include:
- flexible modeling and refinement of timing accuracy according to customized user needs;
- reuse of untimed models to reduce time-to-market of SoC products;
- ability to plug different timing models into the same untimed model;
- dynamic switch to turn timing on/off in a given model;
- legacy management of reusing cycle-accurate models;
- independent, concurrent yet integrated developments between untimed-oriented verification team and timed-oriented architecture team.

4.2   Modeling Approach

To develop a timed model at the transactional level, considerations must be given to the time consumption of two aspects: computation and communication.

The computational delay is the time amount required to perform specific calculations in characterizing a given system behavior or function; whereas the communication delay is the total time consumed in accessing and transferring data or information. The various physical constraints that could bring a significant impact on the system timing behavior such as bus size, bus throughput, or memory size, must also be taken into account during the timed TLM development.

We model the time consumption of a given component in timed TLM through two different tactics:

1. Annotated model
2. Standalone timed model.

4.2.1   Annotated Model

The annotated model is a modeling approach where timing delays are annotated, i.e. inserted, into an untimed model. These annotated delays are the timing information of the *micro-architecture* level, which make the annotated model distinct from the untimed TLM model inserted with functional delay at *architecture* level (as described in Section 3.4).

Here, the delay of each possible set of activation-synchronization in a process is defined based on the control flow of the concerning component. This delay can be modeled with the values of the best, mean, or worst cases. A process could sometimes include very complex control graphs that will consequently entail a large set of timing attributes. If the modeling task becomes too large to handle, a "lazy" approach could probably be adopted by providing only the default conservative values for the unresolved activation-synchronization path. These conservative values constitute the minimum acceptable set of timing constraints that an implementation must comply to.

In general, the annotation approach is well suited if the structure of the untimed model already matches the structure of a micro-architectural model, where annotations will be simple wait statements related to the computation time of a specific functionality. We try to reuse untimed TLM models without any alterations through this approach, although some adaptations could be necessary in certain cases. It is essential to protect the timing annotations with preprocessing directives (e.g. `#ifdef ANNOTATED_MODEL`) in order to select the appropriate execution mode (untimed or annotated) according to user needs.

4.2.2   Standalone Timed Model

The standalone timed model is a different approach where the actual timing behavior is modeled in such a way that delays are computed during the execution of a *standalone* timing model. Our development results have shown that this is applicable on hardware IPs and processor models.

A standalone timed model denotes a detached model incorporated with the timing information. This model is suitable when the structure of the algorithm is very different from the structure of the micro-architecture. Indeed, annotations cannot lead to an accurate timing in such cases. Consider the example of modeling a video application. If modeled at the frame level, only those delays associated with decoding a frame can be annotated. The micro-architecture of the application, on the other hand, allows both the communication and computation to be interleaved.

Conceptually speaking, standalone timed models are high-level analytical timing models *without* functional information. They can be built as traffic generators, which model the channel or interconnect traffic with some timing information.

If the timing behavior of a component depends on its functional behavior, the corresponding standalone timed model can be controlled externally, for instance, by an untimed TLM model. In that case, all the functional events occur during the functional execution of the untimed TLM model must be traced and provided to the standalone timed model. A timing control unit is used to manipulate this information between the untimed TLM and the standalone timed model.

Figure 2-6 gives a better idea about the concept and structure of a standalone timed TLM model combined with an untimed TLM model.

There are two general guidelines to realize the mixed model described above for a given IP. First, develop a purely untimed TLM model describing the functional behavior of the IP regardless of its timing characteristics. Second, develop a timed module in charge of all timing and micro-architecture related information of the IP, without duplicating the functional codes already done in the untimed model. The overall mission of the mixed model is characterized hereafter.

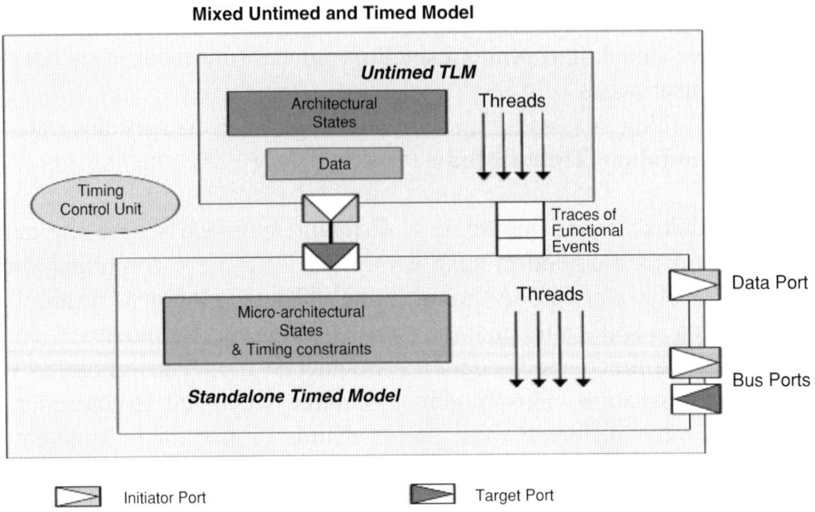

Mixed Untimed and Timed Model

Figure 2-6. Combination of Untimed TLM and Standalone Timed Models

Untimed TLM Model

a) The untimed TLM model executes the pure untimed behavior that will consequently generate or receive transactions through its communication ports. This model must be instrumented for generating traces of functional events, which will trigger certain activities in the timed model.

Standalone Timed Model

a) The standalone timed model implements the mechanism to represent the timing behavior. If the design schedule is too tight to allow developing a very detailed and accurate model, the standalone timed model can be modeled with coarse grains. For a precise implementation, it can be modeled at the micro-architectural level with approximate cycle-accuracy. Standalone timed models are normally controlled by using functional traces generated in the untimed TLM model.

b) The standalone timed model declares communication ports to capture transactions from the untimed model and to insert time delays according to the traces of functional events. Transactions are exchanged through both untimed and timed ports of the timed model. Untimed ports are connected to an untimed communication channel/interconnect while timed ports are connected to a timed channel/interconnect. Details on the model of computation and rules to issue transactions on untimed/timed interconnects are provided in Section 4.3.

The mixed model offers numerous advantages as follows:

- Concurrent development of functional and timing models facilitated by the clear distinction between their modeling strategies.
- Multiple timing scenarios ranging from high-level to very accurate low timing level can be defined, and they can coexist for a unique functional model.
- Untimed models are reusable as the golden reference for functional verification without modifications.
- Optimized speed granted by the dynamic switching between untimed and timed models at the simulation run time.
- Mixed simulations involving timed and RTL models are feasible.
- Architecture and micro-architecture teams can work concurrently on different but complementary models

4.3 Model of Computation

4.3.1 Inter-Execution of Untimed and Timed Models

The working concept of the timed TLM can be pictured as an inter-execution of untimed TLM and standalone timed TLM models. Figure 2-7 illustrates the simulation timelines representing the activities of a process execution in the timed TLM.

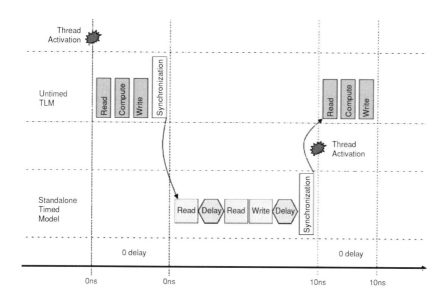

Figure 2-7 Inter-Execution of Untimed and Timed Models

Note that the functional behavior of the untimed model is executed until it reaches a synchronization point. The execution is then passed to the standalone timed model. The timing model will start simulating the delays associated to the functional parts that have just been executed earlier. Meanwhile, time delays of communications and computations are simulated in the timing model as well. Once all of the relevant delays are simulated, the untimed model will resume its execution until the end of its simulation.

The "inter-execution" of untimed and timed models is permissible as long as the untimed model is *fully* modeled using explicit system synchronizations. In this condition, read/write operations are generated only when the data is ready within a stable system. Let us zoom in on the details of such inter-executing mechanism by considering the platform depicted in Figure 2-8. The initiator IP is the master while the target IP is the slave.

The untimed platform is composed of:
- the untimed model of the initiator (I);
- the untimed model of the target (S);
- an untimed communication channel (C).

The bindings for the untimed platform are as follows:
- the initiator port of I is connected to the target port of C;
- the initiator port of C is connected to the target port of S.

In addition, the following modules are instantiated in the platform to support the "inter-execution":
- the standalone timed model of the initiator (TI);
- the standalone timed model of the target (TS);
- a timed communication channel (TC)[1].

---

[1]   Timed channel can be hierarchical to represent the internal topology of the interconnect.

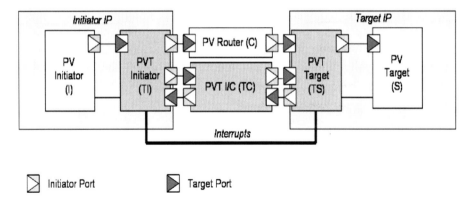

Figure 2-8. Mechanical Structure of Inter-Execution

The bindings related to the inter-execution are as follows:
- the initiator port of I is connected to the untimed target port of TI;
- the untimed initiator port of TI is connected to the target port of C;
- the timed initiator port of TI is connected to the target port of TC;
- the initiator port of TC is connected to the timed target port of TS;
- the initiator port of C is connected to the untimed target port of TS;
- the untimed initiator port of TS is connected to the target port of S.

All sorts of transactional accesses are set off from the initiator to the target through the initiator port; and the functional information is passed to the standalone timed model through the appropriate data structure.

Referring to Figure 2-8, *TI* traps transactions issued by *I*. When *I* meets a synchronization point, the standalone timed model *TI* will start its execution. It computes all of the necessary delays as modeled in the timing model of the micro-architecture, and it issues transactions. As *I* may have generated transactions at a high level of abstraction (e.g. frame), *TI* will generate the appropriate number of transactions from the micro-architectural point of view (e.g. pixel). *TI* may also reorder the transactions to represent read and write interleaves in cases like pipeline.

The overall communication mechanism is as follows:
1. Transactions are issued by *TI* on *C*.
2. Transactions are received by *TS* from *C*. A careful analysis is diagnosed on the transactional access to identify its nature. Depending on the nature of the access, *TS* will handle the transaction accordingly. There are two kinds of accesses:
   a) *insensitive access* - no impact on the IP synchronization scheme.
   b) *sensitive access* - leave impact on the IP synchronization scheme.

For an insensitive access, the simulation continues directly in *TS* for any potential computational time delays associated with the transaction. Indeed, the TLM transaction is propagated "in advance" compared to the actual event occurrence in the silicon. Such advance is permissible on condition that the synchronization scheme can prevent the system consistency from being corrupted by the access. For analysis purposes, the corresponding communication delay from the initiator to the target is passed through the timed channel *TC*, although they will be ignored by the target for the simulation.

For a sensitive access, on the other hand, the transaction emitted from *TI* is rejected by *TS*. Early accesses are not granted in this case because certain behavior could be triggered earlier than what it should be. The adverse consequence will be the undesirable system inconsistency. To prevent this from occurring, *TI* must re-generate the transaction by transferring it through the timed channel, *TC*, in order to include the related communication time delay. The transaction will now be received and accepted by *TS* with the correct time granularity at the right timing. Then, the access will be re-generated by TI on C to actually read/write the data.

Any computational time delay closely related to the initiator or target IP is managed locally by the timed models of the respective IP. Asynchronous events such as interrupts are handled at every single activation boundary. Fine-tuned behavior can be obtained in using pseudo synchronization points as described in Section 4.3.3.

4.3.2   Discussion on Standalone Timed Model Techniques

The standalone timed model is a technique implying a strict compliance with the modeling rules discussed earlier to ensure *no* micro-architectural timing information is implemented in the untimed model. The key advantage is the very neat separation of functional untimed models from micro-architectural timing representations. Thus, it is straightforward to develop several standalone timed models for a given functional model, which allow investigating several micro-architecture scenarios.

With such techniques, the sequence of communication and computation delays may not correspond to the associated functional sequence (while they usually do). For example, an untimed model may grab a full image to process it in one-shot while a timed model would process the data accesses and computations as interleaves. In addition, communication and computation delays can be interleaved in various manners, which could probably be different from its sequence of functional behavior too, e.g. pipeline characteristics. Compared to the functional model, validating the standalone timed model should be handled more carefully to ensure that no error is inserted.

Since the functional and timing information are clearly separated between untimed and standalone timed models, it is possible to couple untimed models with traffic generators. Traffic generators connected to the timed interconnect can act as standalone timed models. The untimed model drives the traffic generator, which is not aware of the functionality but able to generate meaningful bus sequences on the interconnect. This method is particularly useful when traffic generators are developed before transactional models, with the intention of reusing both of them in the future.

4.3.3 Pseudo Synchronization Points

Based on the principles of the system synchronization described so far for the untimed TLM, asynchronous events such as interrupts are perceived only at the activation "boundaries" of the untimed TLM. This is due to the synchronization mechanism coupled to a non-preemptive simulation kernel.

As a process will suspend only on explicit synchronization points, no other processes can execute in the background. While this is not an issue for purely untimed models, it becomes a concern when mixing untimed and standalone timed models. Indeed, asynchronous events may occur too late during the suspended phase of a thread under certain circumstances. Consequently, they may not be caught at the appropriate time.

To handle this problem, finer-grain *pseudo* synchronization points are defined in untimed TLM models. These false synchronization points behave as if many pre-emption points appear more frequently to check for asynchronous event occurrences. They enable timed TLM threads to manage incoming asynchronous events such as those for memory accesses in between synchronization points.

4.3.4 Absolute Micro-Architectural Features

Most of the features for a given system can be modeled as a pure functional model, and can be further refined as a timing model. Certain features, however, are not represented in a pure functional model because they are not relevant at that level of abstraction.

Modeling engineers should be aware of some complex micro-architecture blocks that might be added at the micro-architectural level to optimize the (timing) performance. While such blocks have no relevance to the functional level, it becomes compulsory to model them in a standalone timed model. The reasons are that these blocks definitely related to the micro-architectural information of the system, and they have known impact on the system performance.

TLM can manage such features by integrating the micro-architectural information as well as the related behavior into the timed module. An excellent example to illustrate this idea is the modeling of memory cache.

By definition, a cache is an implementation to improve the performance of the real system. It is not required to be included in the simulation to verify the functional correctness of the design. For this reason, a cache should not be conceived as an architectural model. What we wish to observe in the simulation is the actual traffic of cache activities on the channel for collecting its actual timing figures.

Therefore, the cache needs to be modeled accurately for its traffic and timing changes in the timed simulation as the micro-architectural model. The timed model of the memory cache includes not only timing information, but also some code pieces that reflect the cache effect on the data amount generated onto the channel.

The same approach applies to the reuse of an instruction-accurate ISS in a timed platform. The modeling of the pipeline and cache features as micro-architectural timed models is compulsory to obtain accurate timing figures.

5.   ADVANTAGES OF TLM

Amongst the abundant endeavors proposing modeling techniques at higher abstraction level, TLM has managed to sail its way through to offer a promising solution to SoC industry. As a reliable methodology that can rapidly improve the design productivity, TLM confronts the SoC design bottlenecks in complexity and time pressure through three axes:
1. Early software development.
2. Architecture analysis.
3. Functional verification.

- **Early Software Development**
Software development activities, especially debugging and validation, will have effect only if the software could be executed on its target platform.

Conventionally, a physical prototype such as emulator or FPGA board prototype is considered as the starting point of software development. The downside of this approach is obviously the late availability of such starting point too close to the end of the hardware development. Not only is the time a hindrance, any hardware issues revealed by the software execution at this stage will be too costly to fix as well.

The hardware/software co-verification could of course start executing the software earlier on the target hardware platform. But then again, it still needs to wait quite long for RTL hardware models before running anything.

Rather different from the two approaches mentioned above, TLM SoC platform can be developed right after the delivery of system specifications. The target platform is therefore available for the software development much earlier in the SoC design cycle. In other words, the software development is now conducted in parallel with the lengthy hardware development, i.e. a veritable concurrent hardware/software design is attained.

With the "contract" of TLM platform signed between them, both software and hardware teams cooperate in an independent but converging manner. Software developers regard TLM platform as the reference to run their codes while hardware designers consider it as the golden reference for their RTL design.

In general, software developers employ TLM platform for two kinds of software development:

a) functional software development using untimed TLM;

b) optimized software development using timed TLM.

The greatest advantage of having early software development based on TLM platform is the reduced time-to-market of SoC products through concurrent hardware/software design.

• Architecture Analysis

To increase the chances of first-time silicon success, a system must be thoroughly controlled at each step of the design flow against the real-time constraints stated in the initial system definition. An architecture exploration allowing system performance analysis and verification will fulfill this requirement. The timing information is often essential in such analysis.

System architects and RTL designers seek constantly a better solution for the architecture exploration at an earlier SoC design phase. For this, TLM offers a favorable approach by providing the possibility to explore a system architecture shortly after the system specification is completed. Depending on the user needs, either the untimed TLM inserted with functional delay or the timed TLM can be used for this purpose.

Through an earlier architecture analysis, any system optimization or modification could be handled in time- and cost-efficient way. Besides, it helps to improve the design consistency between hardware and software teams since they are both founded on the same TLM architectural model.

• Functional Verification

Functional verification is intended for assuring the compliance of a given component or system implementation with its corresponding functional specification. RTL models of the design-under-test are analyzed in a functional verification environment by various test scenarios. These test scenarios are developed by verification engineers referring to the paper specification. Most of the time, the engineers need to "manually" determine the expected results of each scenario.

In fact, TLM is the actual functional specification of a component or system. More precisely, TLM is the executable specification of a given design that captures the intended behavior perceived by end-users, i.e. architectural view; but not the implementation details of micro-architectural view. Thus, TLM can replace the manual process undertaken by verification engineers to generate the expected results of test scenarios as the golden reference for functional verification.

Not only is TLM platform used for developing the reference output of test scenarios, it is also reused to conduct functional verification of RTL models with the same test scenarios. The outcomes of the RTL functional verification will be compared to the reference output generated by TLM for analyzing and verifying the design behavior.

As a result, TLM can really save the verification team a huge amount of working time. In addition, it aligns their job constancy with those of software and hardware design teams through referring to the same TLM platform.

6.   CONCLUSION

Concisely, TLM plays the role as the *unique* reference for different teams all the way through the SoC design cycle. Such idea of centralized reference is depicted in Figure 2-9.

Not only is TLM a reliable methodology to face SoC design bottlenecks, it is essentially the single reference that puts into effect a "contract" among the different teams to achieve three durable objectives:

- Work consistency across various teams.
- Rationalization of modeling efforts.
- Cross-team communication and interaction.

In conclusion, the ultimate goal of TLM is leading the SoC industry to a cost- and time-efficient SoC project management in the long run.

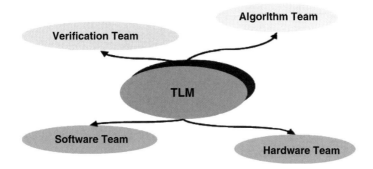

Figure 2-9. TLM as Unique Reference Model

REFERENCES

[1] A. Haverinen, M. Leclercq, N. Weyrich, and D. Wingard, "SystemC-based SoC Communication Modeling for the OCP Protocol," [Online document] 2002 Oct 14 (V 1.0), [cited 2004 Nov 5], Available at HTTP: http://www.ocpip.org/socket/whitepapers/

[2] J. Gerlach and W. Rosenstiel, "System Level Design Using the SystemC Modeling Platform," in Proc. of the Forum on Specification & Design Languages (FDL'00), 2000.

[3] L. Semeria and A. Ghosh, "Methodology for Hardware/Software Co-verification in C/C++," in Proc. of the High-Level Design Validation and Test Workshop (HLDVT'99), 1999.

[4] A. Fin, F. Fummi, M. Martignano, and M. Signoretto, "SystemC: A Homogenous Environment to Test Embedded Systems," in Proc. of the IEEE International Symposium on Hardware/Software Co-design (CODES'01), 2001.

Chapter 3

TLM MODELING TECHNIQUES
Based on SystemC

Laurent Maillet-Contoz and Jean-Philippe Strassen
STMicroelectronics, France

Abstract: The TLM concept and methodology are attainable through an implementation founded on the appropriate system level modeling language. Among the abundant choices of system level languages, we have adopted SystemC as our modeling vector at the transactional level for SoCs. This chapter pulls together our development work to date as a concise illustration of the TLM modeling techniques with a particular focus on SoC communication.

Key words: modeling environment; system level modeling language; SystemC; modeling API; layered approach; core TLM interface; TLM protocol; TLM IP; transaction; initiator; target; interconnect.

1.   INTRODUCTION

After discussing extensively the concept and the methodology of TLM in Chapter 2, the current chapter will deliberate on the techniques employed to support modeling of communications based on the TLM methodology. The modeling of IP behavior at abstraction levels defined in Chapter 2 is not in the scope of this chapter.

Following our research based upon SystemC 2.0, a good understanding of the TLM abstract level was acquired as explained in Chapter 2. Underpinned by this comprehension, we succeeded to develop our own TLM interface [1]. Our development results, demonstrating a good level of maturity, were contributed to the OSCI TLM Working Group, as a significant part for the first OSCI TLM standard [2] delivered together with SystemC 2.1. This chapter discusses the TLM modeling techniques that we have been developing, including our most recent development work in line with the official OSCI TLM standard.

F. Ghenassia (ed.), Transaction Level Modeling with SystemC, 57-94.

To begin with, the modeling environment is introduced in terms of its modeling languages, requirements, and infrastructure. Once familiar with the environment, it is time to learn more about the Application Programming Interface (API) of TLM modeling. Advanced descriptions cover modeling techniques for TLM initiator, target, and interconnect modules. Finally, examples of TLM systems are illustrated, followed by a summary of the chapter.

Targeted readers for this chapter are modeling engineers interested in understanding the TLM internal view from the implementation perspective. For this reason, readers should already have their first exposure to C++ and SystemC constructs before tackling this chapter.

2.   MODELING ENVIRONMENT

The TLM modeling environment is a particular setting wherein TLM models are implemented. An introductory section briefly discusses a list of high-level modeling languages, which are potential candidates to support TLM methodology. Requirements for adopting a given high-level modeling language in the semiconductor industry are also highlighted. Among all the potential candidates, we have chosen SystemC as the implementation language for TLM, covering the development environment and modeling infrastructure. Explanations will be given to justify our choice of SystemC.

2.1   System Level Modeling Languages

2.1.1   Brief Overview

Pressurized by the ever-increasing system complexity, electronic system designers have been struggling hard for years trying to raise the level of abstraction. Since the early 1990s, several proposals of modeling languages for "system-level" design were initiated. These proposed languages can be categorized into three families as follows:

1. *Hardware-oriented Languages*. Most common examples are VHDL or Verilog. As such, these languages offer modeling primitives for hardware design. They are however inconvenient for describing software parts of a SoC.

2. *General-purpose Programming Languages*. The most representative example of this family is C-language. Such languages are suggested from the perspective of software. They rarely offer a full support for

hardware data types and constructs such as concurrency. Again, this family will not be a good choice to describe a complete system.

3. *Proprietary C-based Languages*. These are some sorts of hybrid languages with specific constructs to model SoCs. They are either costly proprietary languages sold by CAD vendors or limited versions reserved for academic purposes. A handful of examples include HardwareC [3] and SpecC [4].

Despite their initiatives, some shortcomings of these modeling languages have hindered their attempts to raise the level of abstraction. The first two families employ heterogeneous co-simulations, which are often inefficient in terms of simulation speed. Proprietary languages in the third family oblige users to obtain a complete tool suite, including compilers and debuggers, for modeling a system. Listed below are disadvantages due to such obligations:

1. Long learning curve for mastering each specific tool suite.
2. Difficult to exchange models between different teams because of license issues and tool suite installations.
3. High cost owing to license fees required for model development and simulation.
4. Uncertainty of support for languages coming from research projects.

2.1.2   Requirements for Industrial Adoption

An appropriate candidate of system level modeling language must fulfill certain conditions required by the semiconductor industry. Collected below are the requirements for the industrial adoption of such languages:

1. Single language for hardware and software modeling in order to achieve high simulation speed.
2. Concurrency for modeling various processes running in parallel within a system.
3. Reactivity for modeling reactive systems sensitive to events or signal changes.
4. Distributiveness for allocating a system simulation on different workstations so that its sub-systems can be simulated in parallel for reaching higher simulation speed.
5. Timing for modeling the system micro-architecture information whenever appropriate and necessary.
6. Data types for specific hardware and software modeling requirements.
7. Tool support from a wide variety of CAD tool vendors.
8. Transparent model sharing between different teams or even companies.
9. Short learning curve for mastering the language.

The modeling languages introduced in the previous section fail to prove themselves in fulfilling all of these requirements for industrial adoption. They only support part of the technical requirements or use a business model with very restrictive constraints for model development and exchange.

2.2   Our Journey to SystemC

After investing some efforts in trying out several proprietary languages, we started our research work on other languages with high potential to raise the level of abstraction for system level design.

- *C/C++ Programming Language*
The first attempt was made on pure C/C++ programming language. To model a complete system in C/C++, a set of primitives must be employed in order to implement the necessary hardware-related data types. We were bound to implement our own simulation kernel, which was able to suspend and resume POSIX[1] threads that modeled concurrent processes defined in a system. In fact, the major drawback of this approach was exactly the implementation of such simulation kernel that was too specific for a given application. Whenever there was a change in the number of processes, the simulation kernel must be updated. System models and simulation engines were nevertheless tightly coupled. The recurrent changes of the simulation kernel made it extremely difficult to identify and fix bugs in the C/C++ implementation.

- *Synchronous Languages*
Slightly wedged by the first attempt, we then quickly shifted our effort to synchronous languages such as Esterel [5] or Lustre. The objective of experimenting Esterel was to understand how a synchronous language with well-defined semantics and concurrency support could really benefit circuit modeling at system level. Appealing advantages were natural support for concurrency as well as strong connection to methods and tools for formal verification. Unfortunately, significant downsides did exist. The learning curve of Esterel was unusually long due to the "cultural shock" that users must overcome in order to be acquainted with such novel paradigms. Our research concluded that synchronous languages were not appropriate entry points for system level design. Today, we seriously consider them as well-founded intermediate representations for applying formal methods, which can be automatically generated from another description. A research project

---

[1]   Portable Operating System for unIX (OS, IEEE 1003, ISO 9945, PASC, UNIX).

is currently studying the role of Lustre in formal verification. Results obtained to date in this research are presented in Chapter 5.

- *System Verilog*

Our next brief stop was System Verilog [6], an extension to the hardware description language, Verilog. This was a high-level language specifically oriented to system modeling and verification. It supported linking to externally defined C functions but not the C++ coding styles. Since it was an extension to Verilog, System Verilog could naturally handle clock-based modeling without much difficulty. However, it reached some limitations in the transactional level modeling. The most obvious problem was that System Verilog was too close to a hardware-based modeling language. It lacked certain capabilities to handle some aspects of higher-level modeling, for instance, abstract data types were not well supported. Furthermore, users of System Verilog were typically those "extended" from Verilog, i.e. they might tend to model something too close to RTL models.

- *SystemC*

After a series of different attempts, we finally reached the long sought-for destination of system level design: SystemC [7]. With the advent of SystemC 2.0, a hardware/software modeling foundation based on a *single* language was no more a dream. SystemC adopts an object-oriented approach that builds a set of classes on top of C++. The essence of SystemC lies in the availability of hardware primitives together with a simulation kernel. With such features, SystemC is able to support multiple abstraction levels and refinement capabilities ranging from high-level functional models to low-level timed, cycle-accurate, and RTL models. Various means are provided to represent communication protocols and channels in high-level modeling. In addition, a help library is made available to assist users with signal and timing details in low-level modeling. SystemC holds all of the C++ operator overloading and pointer capabilities. Therefore, software engineers should feel very comfortable to work with SystemC where the job is mostly done in C++. Since it is a C++ based approach, SystemC offers debugging abilities using classical debuggers such as GDB.

All through our research efforts, we have concluded that an efficient hardware/software system modeling must entail:

1. support for hardware and software primitives;
2. availability of a simulation kernel;
3. dependence on standard C++ compilers.

SystemC, respecting all requirements above, has come into prominence as the most appropriate solution for raising the abstraction level to accommodate system level design. We have therefore opted for SystemC as our modeling vector for SoCs at the transactional level. After implementing our first TLM interface based on SystemC 2.0, we are currently aligning our development work in accordance with the first OSCI TLM standard and SystemC 2.1.

2.3   Development Environment

TLM development environment should support modeling engineers with high efficiency in the development, debugging, and integration of TLM SoC models. We strongly recommend an environment that is independent of any proprietary languages or tools. If this can be respected, the minimum requested tool suite will simply comprise a text editor, a C++ compiler, and a debugger. Modeling engineers can of course choose a more user-friendly environment such as an integrated environment including an editor, a compilation chain, and a debugger.

TLM models developed independent of any tool grants the capacity of recompiling codes on different workstations, using different compilers for code compilations, and running codes on different operating systems. Indeed, such capacities help to verify the robustness of TLM models. Different workstations, compilers, or operating systems have different behavior and features. As a result, TLM models can be verified in various manners and thus fewer characteristics will be untested or missed. This is a real advantage to improve the verification process of SoC.

To gain higher efficiency, TLM models are designed in such a way that any EDA tools that support SystemC can exploit them. Among a wide variety of EDA tool features, listed below are essential to improve platform assembly and debugging:

1. *Automatic assembly of system netlist.* In a component-based approach, models are usually available as off-the-shelf components with well-defined interfaces. Instantiating and binding a huge number of such components at the top level is a painful and error-prone task. This tricky situation can be avoided by using generic descriptions based on the SPIRIT[2] standard that will be discussed in Chapter 7.

2. *System level debugger.* General debuggers such as GDB offer a limited support for multi-thread executions. Such debuggers are typically competent in *micro debugging* but quite restrictive in system level *macro debugging*. A SystemC-based TLM simulation counts largely on synchronization points to trigger switching from one process to another. It is therefore very useful to have a system level

---

[2]   Structure for Packaging, Integrating and Re-using IP within Tool-flows.

debugger that provides advanced features such as stepping through the global execution, visualization of process activations and their dependencies on system events, browsing of design hierarchy, definition of instance-related breakpoints, etc.

3. *Transactional visualization.* Observing and analyzing transactions of a system simulation is another key point to assist model developers and users in understanding dynamic system features. In addition, it serves as a handy visual aid for debugging a system.

2.4   Modeling Infrastructure

Developing new models at TLM abstraction signifies the first step towards system level design and verification flow. Naturally, the reusability of such models becomes the key factor for adopting the TLM approach.

The TLM modeling infrastructure has therefore been developed to help increasing model reuse. This infrastructure is in charge of source code organization and model library management. Users are encouraged to stick to the same organization suggested by the infrastructure in creating their models, which helps to avoid code duplication and facilitate model reuse. A brief description of TLM modeling infrastructure follows.

TLM modeling infrastructure defines SoC design as a project that refers to IP models. Each project is hierarchically organized in several directories listed below:

- *Component.* Contain all TLM component models implemented for a given project. Each component holds its own directory, which can be hierarchically organized and may have inter-project dependencies.

- *Platform.* Represent the location where components and/or test benches are instantiated for creating the top netlist of a design.

- *Software.* Keep embedded software codes to be run on the TLM simulation platform of a given project.

- *Devkit.* Contain simulation facilities such as transaction recorders or helper functions that are not directly parts of TLM models, but useful for the project in certain perspectives or context.

- *External.* Store definitions of external libraries and tools required by the TLM simulation. These definitions may not strictly be related to TLM models. For example, a library containing a processor model is referred by a TLM platform to instantiate an ISS yet it remains an external item.

- *Protocol.* Collect communication protocols that are implemented on top of TLM interfaces to model specificities of real interconnects like buses or network-on-chip.

A root project provides a starter kit comprising TLM development kit definitions, makefile generators as well as commonly used protocols, components, and platforms. Through declarative dependency files, a given SoC project can refer to any components, devkits, or protocols defined in another TLM project. Therefore, previously defined parts can easily be reused in the platform of any projects.

Regarding the TLM code-build, makefile generators provide facilities to generate an object sub-tree automatically. Such object sub-tree is able to handle specificities of operating systems, compiler versions for generating TLM platforms, and SystemC kernel versions (OSCI[3] or provided by a third party).

3.   MODELING API

The current section gives an in-depth description of the modeling API at the transactional level. The discussion will begin with the overview of the layered approach, which separates a functional IP model clearly from its communication interfaces that exchange data with other models.

3.1   Layered Approach

The TLM layered approach offers high-level primitives tailored for the specific modeling needs of modeling engineers. In many cases of system modeling, several communication protocols of different semantics and content are required. Some cases may only require a simple communication protocol that requests a point-to-point connection to pass data from an initiator to a target. On the other hand, certain cases may need a more complex protocol that supports the following features:

1. complex data structures with address, data, byte-enable information;
2. transaction routing onto the communication medium;
3. score-boarding capabilities for verification purposes.

To address such varying modeling requirements, the TLM layered approach has defined three complementary layers listed below:

---

[3]   Open SystemC Initiative.

1. Core TLM interface layer.
2. Protocol layer.
3. IP layer.

These layers are structured in a similar way as those in the network applications. The corresponding implementation details are provided in the coming sections.

3.1.1   Core TLM Interface Layer

The first OSCI TLM standard defines a core TLM interface, upon which TLM protocols and IPs can be developed. According to the SystemC terminology, an interface denotes the specification or convention of some communication services. These services are implemented in communication channels. IP modules, which are aware of the communication services defined by the interface, can then make use of the services via communication ports.

The core TLM interface layer is the foundation of TLM methodology. It defines a transactional level interface, and declares the related transactional level ports accordingly. Indeed, such layer is the *minimum* interface definition required for modeling a SoC at the transactional level. It gets ready a communication API that is capable of transporting a transaction from an initiator to a target module.

3.1.2   Protocol Layer

Various communication protocols can be defined on top of the core TLM interface layer. These protocols rely fully on the core TLM interface to transfer a transaction between two different points in a system. The semantics of the transactional transfer are refined by these protocols in terms of transaction payload and blocking/non-blocking transfer.

As explained in Chapter 2, TLM modules communicate through ports. Initiator ports are defined as a specialization of `sc_port` on the master side, while target ports are defined as a specialization of `sc_export` on the slave side. For each TLM protocol, a set of methods is defined. These methods are also known as convenience functions. They are specified in the form of interfaces in the C++ abstract class. These methods hide the complexity of the core TLM interfaces and thus they *make sense* to the end users. The implementation of these methods must be done in two areas:

- *Initiator Port*. For initiator ports, the implementation of these methods is done once and for all by the protocol developer. It allows translating user-level operations (e.g. read or write) into the appropriate core TLM interface calls.

- *Target Module.* For target modules, the core TLM interface is implemented in a target base class that is done once and for all by the protocol developer. This implementation calls the methods of the protocol interface implemented in each target module. Therefore, the IP developer is responsible for the right implementation of the convenience functions. Since target modules inherit from the protocol interface, the compiler will definitely check for the existence of the implementation of these convenience functions.

The examples of the TLM protocols developed by our efforts include:

1. *TLM_TAC.*
   TAC stands for Transaction Accurate Communication. It specifies a very abstract bus model with a blocking read/write API. It uses the bidirectional blocking core TLM interface, `tlm_transport_if`.

2. *TLM_STBUS.*
   This is a model of the STBus[4] protocol at the packet level. This protocol definition includes the representation of all opcodes and the associated information of STBus. It uses the unidirectional blocking core TLM interface, `tlm_blocking_put_if`, and the non-blocking core TLM interface, `tlm_nonblocking_put_if`.

3. *TLM_SYNCHRO.*
   A TLM protocol particularly developed for purely functional simulations. System synchronizations are modeled without using `sc_signal`. It uses the unidirectional blocking core TLM interface, `tlm_blocking_put_if`.

3.1.3   IP Layer

TLM IPs are modeled on top of a TLM protocol layer as functional modules. Communications between TLM IPs are established through the communication API defined in the protocol interface.

A given TLM IP can instantiate as many ports as required as long as it is in accordance with the underlying protocol. Several ports based on the same protocol could be instantiated by an IP, for instance, a dual port memory IP.

---

[4]   STMicroelectronics proprietary on-chip bus protocol.

Each IP is aware of a set of protocols that it can support. An example of an IP supporting multiple protocols is the protocol converter. This IP receives transactions from a sub-system built on a particular protocol. It then converts the received transactions into another protocol, and reinitiates them to another sub-system as new transactions based on this protocol.

Figure 3-1 illustrates the idea of the TLM layered approach in SystemC.

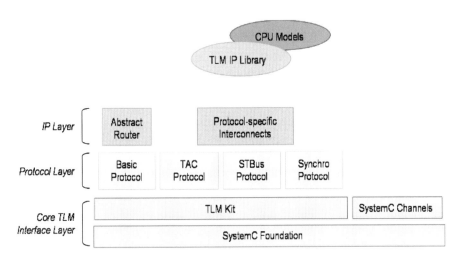

Figure 3-1. TLM Layered Approach in SystemC

3.2 Definition of TLM Protocol

To create a new protocol on top of the core TLM interface layer, some particular definitions are required. First, a suitable core TLM interface must be chosen according to the communication semantics. Then, the protocol between the initiator and the target is defined for two aspects: exchanged information and protocol interface.

3.2.1 Appropriate Selection of Core TLM Interface

The right type of the core TLM interface must be chosen according to the protocol to be implemented. In a given design, one or more protocols can be implemented. Each protocol can either be *unidirectional* or *bidirectional*.

The bidirectional protocol makes use of the blocking bidirectional core TLM interface. Such interface couples the request and the response of a transaction in a blocking call:
- request is a parameter of the function call;
- response is the return value of the function call.

The unidirectional protocol makes use of the blocking and/or non-blocking unidirectional core TLM interface. Such interface is used in cases where:
- initiators send requests and get responses;
- targets get requests and send responses.

The unidirectional interface allows sending or receiving one or more transactions. If it is a blocking interface, the calling process can be suspended in the interface implementation until completion. If it is a non-blocking interface, the calling process cannot be suspended and thus it may fail.

The untimed TLM protocol can be either bidirectional or unidirectional depending on what the role of the protocol is. For example, it is preferably to use the bidirectional interface for an untimed TLM protocol based on the read/write. Certain untimed TLM protocols can use the unidirectional interface, for instance, a protocol that sends the broadcast transaction without waiting for any response.

Since the timed TLM protocol implements timing notions, such as pipe effect, the unidirectional interface is naturally a better choice.

3.2.2   Protocol Definition

There are two important aspects in defining a TLM protocol:
1. Definition of exchanged information.
2. Definition of protocol interface.

- *Definition of Exchanged Information*
  This aspect defines the information to be exchanged between an initiator and a target following a given protocol. Two parts are to be defined:
  a) Request: represent information transmitted by an initiator to a target, e.g. address, data written, byte-enable, etc.
  b) Response: represent information returned by a target to an initiator, e.g. status, data read, error, etc.

- *Definition of Protocol Interface*
  This aspect defines the methods or functions of a given protocol, which is an interface that makes sense to users and hides the complexity of the core TLM interface. Defining the protocol interface has two objectives:
  a) On the initiator side, it provides a user-friendly interface. For example, it defines the read/write (address, data) functions that are in charge of constructing the request structure and invoking the core TLM interface. These functions represent the different operations of the protocol. Depending on the specific information to be transmitted

by the initiator to the target, these functions can have different signatures.

b) On the target side, the core TLM interface is implemented. In this implementation, the protocol interface function is called according to the content of the request structure. For bidirectional interfaces, the implementation of the core TLM interface elaborates and returns the response structure.

3.2.3   Example of TLM_TAC Protocol

This section describes the definition of the TLM protocol through the code examples of our in-house protocol, *TLM_TAC*.

- *Choosing the Appropriate Core TLM Interface*
  The *TLM_TAC* is a protocol defined at the untimed level. It is a very abstract high-level model of the interconnect. Therefore, the details of the communication are not within the scope of this protocol. Moreover, this protocol aims at high-performing simulation. Pulling all these requirements together, the *TLM_TAC* needs the bidirectional blocking core TLM interface.

- *Defining the Exchanged Information*
  The information exchanged in the *TLM_TAC* includes:
  a) Request: address, data, byte-enable, opcode, access mode, tac_metadata, error_reason, etc.
  b) Response: tac_status, data, tac_metadata.

- *Defining the Protocol Interface*
  Five public methods are defined for the *TLM_TAC* protocol interface. Code examples follow.

```
virtual tac_status read (const ADDRESS &address,
                         DATA &data,
                         …) =0;


virtual tac_status write (const ADDRESS &address,
                            const DATA &data, …) =0;


virtual tac_status read_block (const ADDRESS &address,
                          DATA *block_data,
                          const unsigned int number, …) =0;


virtual tac_status write_block (const ADDRESS &address,
```

```
                          const DATA *block_data,
                          const unsigned int number, …)=0;


    virtual tac_status get_target_info (const ADDRESS &address,
                                        tac_metadata &metadata, …)=0;
```

4. INITIATOR MODELING

An initiator is a TLM module that implements one or more processes, which are capable of generating transactions to the interconnect or channel module. The initiator module can be implemented either as the software running on a processor or as the dedicated hardware.

The current section describes the general working mechanism of TLM initiators through the code examples of our in-house protocol, *TLM_TAC*.

For any kinds of implementation, an initiator is always modeled as a SystemC module with the following characteristics:

1. declare processes using SC_THREAD and/or SC_METHOD to model its behavior;
2. instantiate one or more communication ports that are to be bound to communication channels.

The code example quoted hereafter illustrates how to model these two characteristics in a TLM initiator module. Note that this example is a partial version that shows only the essential coding parts.

```
class traffic_generator :
    public sc_module,
    public virtual tlm_module
{
  public :

    // Module ports
    tac_initiator_port<ADDRESS_TYPE,DATA_TYPE> initiator_port;

    SC_HAS_PROCESS(traffic_generator);

    // Constructor
    traffic_generator(sc_module_name module_name);

    // Traffic generator process
    void traffic_generator_func();
};
```

```
traffic_generator::traffic_generator(sc_module_name module_name) :
    sc_module(module_name),
    initiator_port("initiator_port")
{
    // Traffic generator thread
    SC_THREAD(traffic_generator_func);
}
```

As the next code example on the following page will demonstrate, a process within an initiator initiates the communication by:

1. employing the convenient functions of the protocol interface in the initiator port;
2. or creating the objects of tac_request and tac_response, and calling the transport function of tlm_transport_if implemented in the target.

A process that generates transactions should be declared by an SC_THREAD if it encounters some wait statements. From the perspective of modeling, it is recommended to define the granularity of transactional transfers according to the expected accuracy. For example, a video IP modeled at the frame level should have the corresponding transfers completed at the frame level too to avoid irrelevant multiplications of pixel transfers.

If a processor model is integrated, it is desirable to distinguish between regular and debugger accesses. Thus, it will be wise to instantiate a second initiator port reserved for debugger accesses. This port can be connected to either a regular channel or a debugger channel with backdoor accesses.

In certain cases, implementing processes that initiate communications on one or more channels could be necessary. Our suggestion is to implement as many ports as defined in the IP architecture. Model developers should also handle the transaction management in such a way that any premature deletion or edition of the transferred data is avoided. A good practice is to create the thread that generates transactions first, followed by allocating and owning the data structure.

```
traffic_generator :: traffic_generator_func()
{
//code skipped
ADDRESS_TYPE addr = 0x10000;
DATA_TYPE data_write = 0x10, data_read = 0;
tac_status status;
tac_error_reason error_reason;
```

```
//write data at system address addr
status = initiator_port.write(addr, data_write, error_reason);

if (!status.is_ok())
    ERROR_REPORT(2,
                "\t%s: ERROR Write data %d at 0x%x: %s T:%9.9f\n",
                name(),
                (int)data_write,
                (int)addr,
                error_reason.get_reason().c_str(),
                (float)(sc_time_stamp().to_seconds())
                );
else
    DEBUG_REPORT(3,
                "\t%s: Write 0x%x at 0x%x done T:%9.9f\n",
                name(),
                (int)data_write,
                (int)addr,
                (float)(sc_time_stamp().to_seconds())
);


//read at system address addr stored in data_read
status = initiator_port.read(addr,data_read,error_reason);

//code skipped
}
```

5.   TARGET MODELING

5.1   General Guidelines

A target is a TLM module modeled by implementing a SystemC module that takes charge of the IP behavior. The principles of implementing TLM target modules are listed below:

1. Model all registers in a bit-accurate manner.
2. Model IP behavior at the functional level without any micro-architectural details.
3. Implement the protocol interface.

The core TLM interface should also be implemented. To simplify the work of IP developers, the core TLM interface could be implemented in a base class. In target modeling, developers should focus on implementing the communication service and the behavior of the target modules. The next two sections discuss these two aspects in details.

5.2 Interface Modeling

A base class is implemented for each protocol for the sake of simplicity. Such base classes provide the default implementation of the core TLM interface that fits or makes sense for the corresponding protocols.

Consider the *TLM_TAC* protocol that is aware of Read and Write opcodes. The steps of implementing the `transport` function in the *TLM_TAC* base class are as follows:

1. extract opcodes and various parameters from incoming transactions;
2. call the appropriate protocol interface for Read and/or Write, which are to be implemented in the leaf module.

Collected below is the code example that shows how the default implementation of the `transport` interface is accomplished.

```
typedef tac_request <ADDRESS, DATA> request_type
typedef tac_response <DATA> response response_type

template <typename ADDRESS, typename DATA>
  tac_response <DATA>
  tac_slave_base <ADDRESS, DATA> :: transport (const request_type&
                                                            request)
{
  int number = request.get_number();

  response_type response(number);

  switch(request.get_access())
  {
   case READ:
     if (number > 1)
       {
         DEBUG_REPORT(4,
                   "\t%s: Receive read block request T:%9.9f\n",
                   m_slave_name.c_str(),
                   (float)sc_time_stamp().to_seconds());

       response.set_data_ptr(request.get_data_ptr());

       response.set_status(read_block
                       (request.get_address(),
                        response.get_data_ptr(),
                        number,
                        request.get_error_reason(),
```

```
                                    request.get_block_byte_enable(),
                                    request.get_block_byte_enable_period(),
                                    request.get_access_mode(),
                                    request.get_port_id())
                                    );
    }
  else
   {
     DEBUG_REPORT(4,
                     "\t%s: Receive read request T:%9.9f\n",
                     m_slave_name.c_str(),
                     (float)sc_time_stamp().to_seconds());

     response.set_data_ptr(request.get_data_ptr());

     response.set_status(read
                         (request.get_address(),
                          *response.get_data_ptr(),
                          request.get_error_reason(),
                          request.get_byte_enable(),
                          request.get_access_mode(),
                          request.get_port_id())
                          );
   }
break;
case WRITE:

  if (number > 1)
   {
     DEBUG_REPORT(4,
                     "\t%s: Receive write block request T:%9.9f\n",
                     m_slave_name.c_str(),
                     (float)sc_time_stamp().to_seconds());

     response.set_status(write_block
                         (request.get_address(),
                          request.get_data_ptr(),
                          number,
                          request.get_error_reason(),
                          request.get_block_byte_enable(),
                          request.get_block_byte_enable_period(),
                          request.get_access_mode(),
                          request.get_port_id())
                          );
    }
  else
   {
```

```
        DEBUG_REPORT(4,
                    "\t%s: Receive write request T:%9.9f\n",
                    m_slave_name.c_str(),
                    (float)sc_time_stamp().to_seconds());
        response.set_status(write
                            (request.get_address(),
                            request.get_data(),
                            request.get_error_reason(),
                            request.get_byte_enable(),
                            request.get_access_mode(),
                            request.get_port_id())
                            );
        }
    break;
    case GET_TARGET_INFO:
        DEBUG_REPORT(4,
                    "\t%s: Receive get_info request T:%9.9f\n",
                    m_slave_name.c_str(),
                    (float)sc_time_stamp().to_seconds());
        response.set_metadata(request.get_metadata());
        response.set_status(get_target_info
                            (request.get_address(),
                            response.get_metadata(),
                            request.get_error_reason(),
                            request.get_port_id())
                            );
    break;
    default:
        string msg("Unknown TAC access type");
        ERROR_REPORT(2,
                    "\t%s: %s T:%9.9f\n",
                    m_slave_name.c_str(),
                    msg.c_str(),
                    (float)(sc_time_stamp().to_seconds()));
        response.get_status().set_error();
        request.get_error_reason().set_reason(msg);
break;
        }
  return response;
  }
}
```

Through such default implementation of the tlm_transport_if interface in the protocol base classes, model developers must provide an implementation of the protocol interface in target modules. An example of implementing a read method in a simple timer is given hereafter.

```
//timer read access
tac_status
tac_timer :: read(const ADDRESS& address,
                  DATA& data,
                  tac_error_reason& error_reason,
                  const unsigned int byte_enable,
                  const unsigned int mode,
                  const unsigned int target_port_id)
{
   tac_status status;
   switch (address)
   {
     case TIMER_LOAD:
        data = m_timer_load;
        break;
     case TIMER_VALUE:
        data = m_timer_value;
        break;
     case TIMER_CONTROL:
        data = m_timer_control;
        break;
     default:

     ERROR_REPORT
          (2,
          "\t%s: Error, cannot read at address 0x%lx T:%9.9f\n",
          name(),
          address,
          (float)(sc_time_stamp().to_seconds()));
     error_reason.set_reason("Input read address out of range");
     status.set_error();
     return(status);
   }
   status.set_ok();
   return(status);
}
```

The basic mechanism described above provides communication services for those target modules using a single TLM protocol. If this default mechanism is insufficient, the `transport` function defined in the protocol base class must be overridden to provide implementations that are more specific. In such cases, model developers will have to implement the `transport` function in the leaf module.

Target modules of multi-protocol support must instantiate at least a target port and inherit from target base classes supported by each protocol. In addition, these targets have to implement the methods of each protocol interface. Compilers must be able to distinguish the available alternatives according to the signatures of the various functions.

From the perspective of modeling, IP models should not include any system address in order to allow remapping and multiple instantiations of the modules. Registers can be identified by their offset addresses with respect to the beginning addresses of IP modules. For an IP connected to multiple channels of the same protocol, it is unnecessary to instantiate multiple interfaces. Instead, simply use the target port ID to distinguish the different incoming transactions from the channels.

For a complex IP, the manual modeling of accesses to its numerous registers could be a dull and error-prone task. This is more likely to happen if the modeling is not done in a systematic manner. An adverse consequence of such is the run-time error. This particular modeling code piece, however, can be automatically generated if IPs are specified using a strictly regular structure. Such regular structures can be obtained by either parsing specification documents or providing intermediate register representations like those of SPIRIT format. In that case, register structures are loadable from files to provide fully dynamic register representations, or to generate necessary codes to be compiled with IP models in a static approach. The first option of parsing specification documents is certainly more flexible. In spite of this, it has a negative impact on the overall simulation speed. This option must therefore be considered with meticulous care.

A very common question about modeling target IPs is the amount of ports to be instantiated in TLM models. At the transactional level, an IP is normally modeled in an abstract way that does not match its actual number of ports or bus interfaces as in its RTL model. It will not raise any issue as long as the expected behavior can be modeled correctly. For this matter, we recommend instantiating multiple ports:

1. if it is necessary to distinguish the nature of the ports of the incoming transactions, e.g. arbitration purposes;
2. if IPs are connected to several channels that represent several buses on the actual chip.

If neither case above is encountered, a single target port will be sufficient to collect all of the incoming transactions for a target module.

5.3   Behavior Modeling

Once the communication service of a target module is implemented, model developers should proceed to the implementation of the IP behavior. Two aspects must be handled carefully:
1. side effects of writing or reading data into/from IP registers;
2. ability to trigger some potential processes when registers are properly programmed.

An important reminder here: IP behavior must be modeled at the functional level for untimed models without any micro-architectural details. No communication-specific processes are implemented in target modules. The IP behavior can nevertheless be implemented by SC_METHOD or SC_THREAD as in any standard SystemC models. Certain complex IPs with both the slave and master roles, such as a video coder-decoder (CODEC), may also instantiate initiator ports to read or write data from/into memories.

6.   INTERCONNECT MODELING

An interconnect or channel is a structure responsible for establishing communications among TLM modules. It can be modeled in various manners at the transactional level depending on the expected accuracy as well as the model used on the final platform. The minimum features of a TLM interconnect are the abilities to:
1. decode addresses;
2. route a transaction from an initiator to a target module.

This approach is sufficient to model untimed modules for fast functional simulations during the embedded software development. It is obviously insufficient for an architectural analysis. In that case, the topology (e.g. nodes and links between nodes of a network-on-chip) of the interconnect module will have to be modeled along with its arbitration policy and potential delays.

6.1   Interconnect Structure

In the untimed TLM, an interconnect is modeled as a hierarchical channel with communication ports and the implementation of one of the core

TLM interfaces. Depending on the communication purposes, communication ports of a channel can be arranged differently as:

- *Simple Transaction Router*
  If a channel simply routes transactions without considering arbitration, a single initiator port and a single target port will be sufficient. Each of these communication ports can be bound to one or more IP ports. Indeed, an initiator port and a target port of a router are centralized points to receive and transmit transactions collectively from all IP ports involved in the system communication. Figure 3-2(a) illustrates the interconnect structure of a transaction router. However, it could be useful to offer multi-port routers to facilitate an automatic netlist generation (see Chapter 7).

- *Arbiter Channel*
  A sophisticated channel with arbitration ability is typically implemented for cases where the system topology and port notions are required for handling arbitration. In an arbiter channel, an initiator port or a target port will be particularly assigned for each IP port involved in the system communication. The interconnect structure of an arbiter is depicted in Figure 3-2(b).

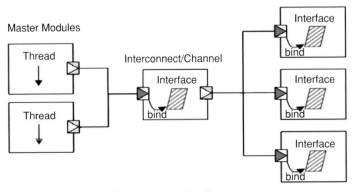

(a) Simple Transaction Router

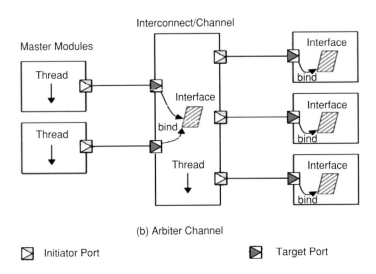

(b) Arbiter Channel

Figure 3-2. Interconnect Structure

On top of having a router implementation for a given channel, it is sometimes useful to have another implementation closer to its RTL model. This can be achieved by having as many communication ports as there are in the real hardware, which is exactly the second case described in Figure 3-2. With both implementations, the channel can either employ a simple routing procedure, or implement an arbitration policy that uses the slave port ID to distinguish incoming transactions. Note that the latter option requires implementing additional processes that may influence the overall simulation speed.

If TLM platforms are applied for an architectural analysis, it will be necessary to provide more details for the interconnect structure. In that case, the TLM interconnect must support all the information that may affect the system performance in terms of timing, bandwidth, arbitration policy, etc. The TLM model of a network-on-chip should instantiate the whole set of modules required to represent its topology, the different paths of the network between initiators and targets, and the different arbitration policies hosted by the interconnect nodes.

6.2   Examples of Interconnect Implementation

6.2.1   Router Implementation

Let us consider the implementation of the simplest case for a router with only an initiator port and a target port. When such a router is instantiated for a given system, it loads the system address map from a file that describes the address ranges associated with each port of all the targeted system IPs. The data structure created through this file loading will serve the transactional routing of the system later on.

From the angle of an initiator, a router is regarded as a target module. Therefore, the core TLM interface is implemented in the router. The minimum implementation of such lies in:

1. correct extraction of the address from the transaction payload;
2. exact information of address map to identify the right target module.

Once the target module is identified, the router will propagate the transaction by invoking the core TLM interface, which is implemented in the target IP and bound to its target port.

Since the semantics of such transactional transfer is a blocking/atomic communication, no process will be needed in the router for transmitting the transactions.

6.2.2   Arbiter Implementation

For the functional part of the embedded software development, a TLM arbiter channel is neither necessary nor recommended. The architecture analysis of a system, on the other hand, will find TLM arbiters of great help.

An arbiter has all the features of a router channel as presented earlier. On top of the address decoding and routing abilities of a router, the arbiter must be able to reorder transactions according to a specific arbitration law.

Two types of implementation are available for enabling arbitration in an interconnect module. First, arbitrate the initiator modules that generate the requests for transmitting transactions. This approach may only have to implement the minimum number of ports in the arbiter channel, which is decided by the arbitration law according to the information located in the initiator ports of the requesting initiators. Second, arbitrate the target modules that attempt to access to the arbiter for passing transactions.

The arbiter must suspend the thread of an incoming transaction despite the nature of the arbitration policy. Such obligation permits other processes to execute, and it allows other potential initiators to generate transactions too. Any pending transaction will be stored in a data structure.

Once all of the potential transactions are expressed, the arbitration mechanism will execute. This mechanism is implemented in another process within the arbiter and is notified by using a delta cycle delay. Consequently, it guarantees that the arbitration will only take place after all the pending requests are received.

When the arbitration process executes, it applies the arbitration law to select the transaction to be served, followed by the address decoding and transaction routing as what a simple router does. As soon as the transaction service completes its job on the target side, the arbitration thread will send an event notification to release the suspended initiator process so that it can continue its execution.

7.   EXAMPLES OF TLM SYSTEM

7.1   Multimedia Platform: MPEG4 CODEC

Our first example is the TLM model of an MPEG4 CODEC based on the TLM_TAC protocol. This multimedia platform will demonstrate how efficient an architectural TLM simulation model can help the embedded software development and interactive debugging.

7.1.1   System Model

The MPEG4 CODEC platform employs a distributed multi-processing architecture with two internal buses and four application-specific processors (ASIP); distributed above which include the embedded software, a general-purpose host processor managing the application-level control as well as seven hardware IP blocks. Consequently, there is a great deal of software parallelism with explicit synchronization elements in the software. Each processing block is dedicated to a particular part of the CODEC algorithm.

The hardware and software are partitioned according to the complexity and flexibility required by the CODEC algorithm. All operators work at the macro-block level, i.e. the video unit for CODEC. Figure 3-3 illustrates the MPEG4 CODEC platform block diagram.

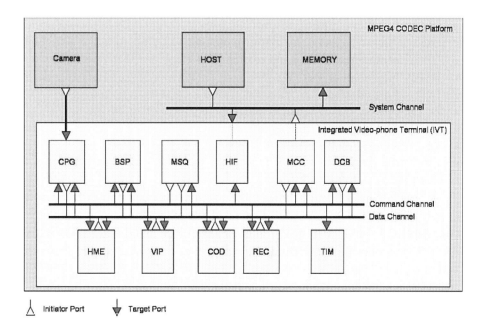

Figure 3-3. MPEG4 CODEC Platform Block Diagram

Listed below are the descriptions of the major CODEC system blocks:

* *Multi-Sequencer (MSQ)*
  This is the RISC processor for video pipeline management. It comprises mainly firmware with a hardware scheduler particularly for fast context switches and process management.
* *Multi-Channel Controller (MCC)*
  MCC is a hardware block with the micro-programmed DMA. It arbitrates all the current requests from the operators and performs I/O with the external memory. Scheduling is done per request where each request is a memory burst with a variable size. An I/O request is considered as a single transaction even if it may take several cycles to execute.
* *VLIW Image Predictor (VIP)*
  This block is a mixture of hardware and firmware for performing the motion compensation. The firmware controls the VIP block while the

hardware handles the processing part with a special instruction set of very long instruction word (VLIW) type. This instruction-set is modeled for the CODEC TLM platform.

- *Encoder (COD)*
  COD is a pure hardware block that performs the difference between the grabbed and predicted image, discrete cosine transformation (DCT), zigzag, quantization, and run/level coding.

Other system blocks on the platform include:
- CPG: camera processing and grabbing
- BSP: bit stream processing
- HIF: host interface
- DCB: display and compositor block
- HME: hierarchical motion estimator
- REC: re-constructor block
- DFO: digital filter

In terms of the system behavior, the external processor posts commands to the mailbox, HIF. The C-programmed MSQ will consider the posted command and take charge of the internal control of CODEC. Consequently, the MSQ activates the different internal hardware or programmable blocks to perform the coding and decoding of the video flow by reading status registers and writing command registers for each block.

All of the operators are pipelined. They communicate with the system memory through a memory controller. This memory controller receives requests from the internal operators and generates transactions onto the top-level bus for accessing to the system memory. It communicates with the internal operators through well-identified input/output registers that contain the values to be stored into and loaded from the memory.

The two internal channels on the platform, command and data channels, serve for different purposes. For those modules that need to give access to their control registers, their ports are bound to the command channel; for those modules that handle data communication, their ports are bound to the data channel. The MSQ is a particular case since it has a double role: it generates read/write operations to control the system behavior, thus its command master port is connected to the command channel; it also initiates transactions to the memory, thus its communication port is bound to the data channel.

7.1.2 Design Choices

The obligation of being able to simulate the embedded software prevents us from developing a pure functional and sequential circuit model. Each computing block, for this reason, is modeled as a SystemC module with its own processes as well as the associated synchronization elements.

The MPEG4 CODEC platform is a multi-processor design composed of C-programmed modules, pure hardware blocks as well as mixed blocks of hardware and programmable operators. On this platform, there are three categories of specific modeling strategies as explained below.

- *Software Block*

The design relies much on the sequential aspects of the codes for MSQ, BSP, HME, and VIP. All firmware is written in C. This enables the codes to be natively compiled on a workstation, and to communicate with the TLM models through an I/O library via simple stubs to call C++ from C. The specific built-in instructions of the processors are modeled as C functions. A part of the C model is used directly for the ROM code generation using retargetable C cross-compilers. An ISS is integrated in the environment to run the cross-compiled application software on the SoC host. The written software remains unchanged despite the nature of its environment, be it TLM or RTL simulations, emulation or application board.

- *Hardware Block*

For each hardware block, a high-level model is written in SystemC. It is a functional, bit-true model with the representation of memory transactions. The internal block structure is not characterized. Instead, the input/output of the block, the system synchronization, and the internal computation at the functional level are modeled. The completed SystemC model is used as the reference for the RTL validation.

Consider COD, the hardware pipelined with five operations: Delta, DCT, Zigzag, Quantization, and Run/Level. A FIFO is inserted among all of these operations. The computation is controlled by a pipelined finite state machine (FSM) while a DMA manages the inputs and outputs. Indeed, the RTL model is fully representative of this architecture. Such a model normally requires at least 3-men-months effort for a senior designer. On the other hand, the corresponding transactional model consists in the input data acquisition (grabbed and predicted blocks in our example), computational results by a C function, and the resulted output. The gain is not only an easier and faster modeling, but also a much greater simulation speed. Furthermore, the required effort is only about 1-man-week.

- *Mixed Hardware/Software Block*

Mixed hardware/software blocks are modeled as a combination of the two previous categories.

7.1.3   System Integration Strategy

Some features of the modeling environment applied in the MPEG4 CODEC platform are briefly discussed in this section. The discussion will only focus on the fundamental aspects of our approach.

The system synchronization of the platform is dual. First, the MSQ is in charge of the global control of the platform. It is responsible for activating the coding/decoding tasks according to the current system status. A task will be executed only if the relevant data is available, otherwise it will simply be suspended. Several tasks may be activated at the same time through the internal pipeline of operations. The system synchronization is achieved by writing/reading into/from the command and status registers of the different operators. Second, data exchanges between the operators and the memory are blocking operations. The platform synchronization scheme ensures that an operator will resume its computation only when the previous transaction is completed from the system point of view.

Data exchanges are modeled with arbitrary sizes in the platform with respect to the semantics of the system data exchange. For instance, transactions between the camera and grabber are line-by-line image transfers while transactions between MSQ and other operators are 32-bits wide.

7.1.4   Experiment Results

The performance figures obtained for the MPEG4 CODEC platform in terms of its code size and simulation speed on different complementary environments are depicted in Figure 3-4.

The modeling choices of IPs can bring a significant gain in terms of the model size. The code sizes of the RTL, TLM, and Firmware models of the MPEG4 CODEC platform are compared in Figure 3-4. Note that TLM models are about 10 times *smaller* than RTL models. Obviously, TLM models are easier to write and consequently faster to simulate. Figure 3-4 also shows that TLM models manage to simulate 1400 times *faster* than RTL models on a SUN-Ultra10 workstation of 330MHz speed and 256MB memory. Just consider the typical job of image coding to get the feel of this speed-up in simulation: RTL models simulate a coded image in an hour but SystemC TLM models only need 2.5 second to do the same job!

For the emulation of the MPEG4 CODEC platform, an ad-hoc co-emulation transactional interface had been implemented through the C API before the Accellera SCE-MI interface [8] was made available on the

Mentor Celaro hardware emulator. This interface provided both controllability and observability of the emulated design over the clock control, the memory and register load/dump as well as the built-in programmable logic analyzer. For the hardware, synthesizable models were developed for the camera, the memory and several RTL transactors [8] translating a data packet from a transaction into a bus-cycle-accurate data exchange. For the software, the TLM model of the host was extended for hardware debugging and performance evaluation purposes. Such co-emulation required 35 second for processing each coded image, i.e. a system clock at about 40-60 kHz was necessary. This speed was more than 30 times faster than a cycle-based co-emulation; it enabled running the software developed for the TLM platform without any external or synthesizable CPU core aimed at the host modeling.

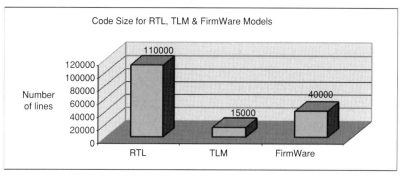

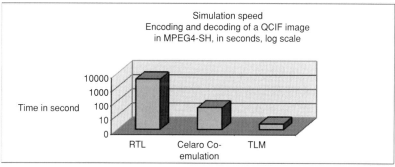

Figure 3-4. Performance Figures of MPEG4 CODEC Platform

Our experiment illustrated that the MPEG4 CODEC TLM platform was able to run a significant test-bench of 50 images in a couple of minutes on a SUN-Ultra10 workstation with full source-level debugging facilities. Based on such performance results, we concluded with confidence that the MPEG4

CODEC TLM platform was very well suited as a base platform for the embedded software development.

7.2   PWP Sub-system: DMAC PL080

Our second example is the TLM model of the PrimeCell Direct Memory Access Controller (DMAC PL080), which represents a sub-system on the ARM PrimeXsys Wireless Platform (PWP) [9].

The focal discussion point of this example is the necessary modeling decisions and trade-offs associated with the development of TLM models. The impact of such decisions or trade-offs on the subsequent SoC design activities, particularly SoC performance analysis, will also be discussed extensively.

The 8-channel prioritized DMAC serves as an excellent example to illustrate TLM modeling choices because it requires a parallel modeling style to capture the whole controller behavior without flaw. Furthermore, this controller challenges the TLM modeling with part of its hardware parallelism that is *not* controlled by the software *but* the hardware logic.

7.2.1   TLM Model of DMAC PL080

The main features of the ARM DMAC PL080 are listed below:

- 8 prioritized DMA channels;
- support for word, half-word, and byte transfers;
- 2 master AHB[5] bus ports for larger data throughput on two buses;
- 16 peripheral-controlled interfaces allowing transfers to be controlled by peripherals instead of the DMA controller;
- peripheral-controlled interfaces are controlled by either hardware signals, or the special SoftXReq register bits that are set by some embedded software external of the DMAC;
- support for both little-endian and big-endian transfer modes (the two AHB bus ports can be programmed separately to support each mode);
- support for an AHB slave interface programming the DMAC memory-mapped internal registers.

Based on the specifications above, the TLM model of the DMAC has to model the following interfaces:

- 1 AHB slave port;
- 2 AHB master ports;

---

[5]   Advanced High-performance Bus.

- 16 peripheral-controlled signals of DMACSREQ, DMACBREQ, DMACLBREQ, DMACLSREQ, DMACCLR, DMACTC;
- 3 DMA interrupt signals of DMACINTERROR, DMACINTTC, DMACINTCOMBINE.

The TLM interfaces and the internal behavior of the DMAC are defined once the features listed earlier are thoroughly understood. The modeling of AHB and APB[6] data ports are based on the TLM API while the system synchronization is based on the native simulator mechanism. Both signals and events are available in SystemC. However, it is preferable to use signals in TLM modeling. The reason is that it facilitates the application of TLM models as the golden reference model for RTL verification and co-simulation. Following are the details regarding the modeling choices made for the main features of the DMAC.

- *AHB Slave Port*
  This is an interface for accessing to the DMAC registers. Care must be taken that the *a priori* asynchronous update of the DMAC TLM slave registers by the external software, does not and should not corrupt the internal behavior that depends on these registers. Since SystemC runs with non-preemptive threads, its simulation is controllable in each thread at known preemption points only. Model developers must therefore ensure that preemptions points are carefully defined. Besides, the system synchronization must be correctly modeled to achieve a safe system design and a flawless SystemC simulation.

- *AHB Master Port*
  The two AHB master ports of the DMAC allow simultaneous read and write operations. If a DMAC TLM platform is intended for an architectural analysis, both ports are required in order to provide a realistic traffic generation. Since each of the ports has specific registers accessible by the embedded software, both ports must be modeled as well for a DMAC TLM platform intended only for the software development.

- *Interrupt Signal*
  Three of the interrupt signals in DMAC must be modeled because they represent system events.

- *Register Bank*
  The DMAC registers are modeled as the data members of the class definitions of DMAC TLM modules. All of the TLM DMAC registers

---

[6] Advanced Peripheral Bus.

can be set to the equal size, e.g. to 32 bits. It can be set to the maximum register size or the word size of the host machine (whichever has the larger value will be chosen).

In TLM modeling, a general rule to model the internal behavior of an IP is that any deviation from the IP functionality is permissible as long as the functionality remains unchanged from the software point of view. Such modeling abstraction reduces greatly the modeling effort and simulation time of the IP.

The DMAC includes two internal arbiters, one for each bus port. The arbiters implement a priority-based algorithm, meaning that the channel of the highest priority is allowed to transfer data from its pending transactions. Whenever a channel is suspended from transferring data due to unavailable input data, the arbiters will grant access to another channel of lower priority. This feature helps to optimize the data transfer although it is not compulsory in the TLM model for a correct execution of the embedded software. Note that the arbiters cannot simply be annotated with timing information to represent their timing behavior. Instead, a correct interleaving of the data transfers between the active channels has to be modeled for this purpose. The addition of such scheduling scheme refines the arbiter model into a timed TLM model, which is more suitable for the architecture analysis.

7.2.2   Performance Analysis

The simulation results of using the TLM model of DMAC PL080 along with the back-annotated TLM model of a static memory controller are carefully analyzed in this section. This analysis serves for illustrating how an industrial platform (i.e. PWP sub-system) can be modeled in TLM and simulated close to the RTL precision, *but* with a much-advanced availability and faster simulation speed than the conventional RTL approach.

If the DMAC TLM platform is initially conceived for the embedded software development, it must be upgraded to enable the performance analysis activities for several reasons listed below:

1. the bus traffic generated by the processor is highly dependent upon its cache behavior;
2. the multi-layer bus structure includes some arbitration scheme to access shared slave modules;
3. the multi-port memory controllers employ particular arbitration policies to select incoming requests;
4. the wait states of memory accesses have no fixed values.

The following are necessary to build a DMAC TLM platform that is appropriate for conducting the performance analysis:

1. an instruction set simulator (ISS) that includes a cache model for the integrated processor in the platform;
2. the initial TLM model of the DMAC has to be refined into a micro-architectural TLM model as described in the previous section;
3. realistic arbitration policies must be implemented to control accesses to the shared resources on the platform.

Memory latencies on the DMAC platform need to be accurately modeled, as we shall describe here for the static memory controller. The number of wait states observed at a given port depends on the transaction transfers of the previous and current accesses to that port.

If an access is initiated while another transaction is accessing the static memory, then the access will be able to finish its job after a fixed time amount that the first access completes its transfer. This fixed duration depends on the memory area being accessed (i.e. the memory bank) as well as the access direction of the previous and current transaction transfers (i.e. read or write). For instance, a 4-port and 8-bank memory controller has 256 situations where each has a fixed number of wait states observed in the RTL simulation. The proper way to time-annotate the TLM model of this static memory controller is to *overload* the read/write methods of its functional TLM model. In the time-annotated model, read/write methods must be able to call the SystemC wait function with a delay equivalent to the number of cycles for a given situation (i.e. each of the 256 situations is represented by a specific duration of delay).

With such time annotations, the resulted simulation platform is a blend of architectural and micro-architectural TLM models that contain sufficient timing accuracy for conducting the performance analysis. Therefore, the timed TLM simulation of the static memory controller can serve as an adequate cycle count estimate with respect to the RTL simulation. The traces of such performance analyses are demonstrated in Figure 3-5 and Figure 3-6, respectively for RTL and TLM simulations.

The upper part of each figure represents the transactions on the AHB instruction bus (AHB-I) while the lower part represents the transactions on the AHB data bus (AHB-D). Both buses are connected to a single static memory controller (SMC) via separate ports. The data is stored in a flash whereas the instructions are stored in a ROM. Thus, the two AHB buses may need to access to the memory banks concurrently through the SMC. When such a conflict occurs, the instruction-fetch transaction will take longer to finish. This phenomenon is observable in both the RTL and timed TLM simulations. Through implementing the RTL-based back-annotations in the memory controller, TLM accesses manage to have the same durations as those in the RTL simulation *but* with a much higher simulation speed.

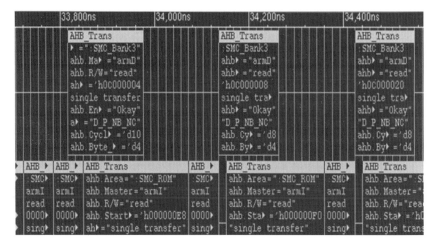

Figure 3-5. RTL Simulation Traces of Performance Analysis

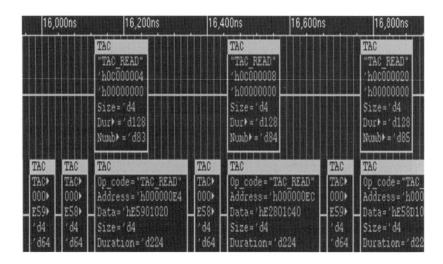

Figure 3-6. TLM Simulation Traces of Performance Analysis

In conclusion, the time-annotated TLM significantly assist architects in estimating the SoC performance through light modeling efforts and the high simulation speed. The greatest advantage of such is the hardware/software co-simulation using modifiable hardware models, and above all, without going through the hassle of modeling the RTL or cycle-accurate platform for the whole SoC.

8. SUMMARY

The fundamental modeling techniques for the TLM approach based on SystemC are gathered in this chapter.

To begin with, a brief overview of the TLM modeling environment is covered in Section 2 ranging from system level languages to modeling environment and infrastructure. Extensive discussions on TLM modeling API are grouped in Section 3. The foundation of the TLM API is introduced as a layered structure that hides nicely the modeling complexity from the end users. It implements the core TLM interface as the minimum interface definition for the transactional level modeling as a communication API to transport a transaction from an initiator to a target. Above this layer, TLM protocols are implemented to refine the semantics of the transactional transfer in terms of transaction payload and blocking/non-blocking transfer. TLM IPs are modeled on top of a TLM protocol layer as functional modules.

Section 4 deals with the initiator modeling. It explains how to create SystemC modules that instantiate initiator ports, and how to get ready processes that implement the IP behavior and generate transactions in a system. The target modeling is elaborated in Section 5. It describes how target modules are modeled by implementing the core TLM interface, either through a default implementation by the protocol base class or using an overloaded core TLM interface implemented locally in the target module itself. Other topics include architectural resource modeling and behavior implementation. An approach to model TLM interconnects is illustrated in Section 6, along with two examples: a simple router and an arbiter.

Before closing the chapter, two practical examples are presented in Section 7 to demonstrate how real SoCs are modeled as TLM systems.

REFERENCES

[1] A. Clouard, K. Jain, F. Ghenassia, L. Maillet-Contoz, and J.P. Strassen, "Using Transactional Level Models in a SoC Design Flow," Chapter 2, <u>SystemC Methodologies and Applications</u>, Ed. W. Müller, W. Rosentiel, J. Ruf, Kluwer Academic Publishers, 2003, pp. 29-63.

[2] A. Rose, S. Swan, J. Pierce, and J.M Fernandez, "OSCI TLM Standard Whitepaper: Transaction Level Modeling in SystemC," [Online document] [cited 2005 April 28], Available at HTTP: http://www.systemc.org

[3] D. Ku and G. De Micheli, "HardwareC: A Language for Hardware Design," in <u>Technical Report CSL-RT-90-419</u>, Computer Systems Laboratory – Stanford University, 1990.

[4]   D.D. Gajski, J.W. Zhu, R. Dömer, A. Gerstlauer, and S.Q. Zhao, <u>SpecC: Specification Language and Methodology</u>, Kluwer Academic Publishers, 2000.

[5]   G. Berry and L. Cosserat, "The ESTEREL Synchronous Programming Language and Its Mathematical Semantics," in <u>Proc. of the Seminar on Concurrency</u>, vol. 197, pp. 389-448, 1984.

[6]   Information available at HTTP: <u>http://www.systemverilog.org</u>

[7]   Information available at HTTP: <u>http://www.systemc.org</u>

[8]   Information available on the website of the Accellera Interfaces Technical Committee at: <u>http://www.eda.org/itc</u>

[9]   Information available at HTTP: <u>http://www.arm.com</u>

Chapter 4

EMBEDDED SOFTWARE DEVELOPMENT
Through The TLM Approach

Eric Paire
STMicroelectronics, France

With special participation of Kshitiz Jain, Marc Harbonne, Maxime Fiandino, and Michel Bruant.

Abstract: Early embedded software development, covering coding, testing, integration and validation, is one of the most important targets of TLM platform methodology. This chapter describes mainly the close relationship between the TLM platform and the software running on it. The description illustrates how the software can benefit greatly from the early TLM platform availability. Reciprocally, hardware developers can also benefit from the early feedback on their design when used by the software developers. The TLM platform can therefore be considered as the meeting point between hardware and software development teams.

Key words: software; Operating Systems; firmware; device drivers; application; protocol stack.

1.   INTRODUCTION

Nowadays, no hardware design of a system-on-chip is worth developing without any software to exercise its functions. The trend of "the smaller the better" in SoC design concept has rapidly pushed the role of software into prominence during SoC hardware design process. While hardware aspects are getting very tough to handle due to the ever-rising SoC complexity, the weight of software aspects becomes more and more important in the overall system to manage new hardware functionalities and to replace certain hardware features.

This chapter highlights the brand-new role of software in conjunction with TLM platforms. It underlines the core idea of how system embedded

F. Ghenassia (ed.), Transaction Level Modeling with SystemC, 95-151.

software and TLM platforms could enhance and enrich each other in their respective missions.

The conventional design approach allows a significant amount of the software being developed, compiled and tested before any strict form of the hardware platform is made available. Only a specific part of software could be developed when the detailed information tightly associated with the hardware is accessible in the form of RTL or emulation platform. This part is usually the toughest and longest to test and debug. Unfortunately, software developers are always bound to wait quite long for such hardware platform in order to validate their development work. This is not only a costly time loss, but also an inefficient cooperation between hardware and software designers for lack of a common development base.

Despite the somewhat opposed design philosophies between hardware and software fellows, current SoC complexity is urging these two worlds to work together in a new way leading to concurrent hardware/software design. Time-to-market reduction and cost saving will be the successful culmination of such parallel hardware/software design.

The idea of hardware/software co-design and co-implementation can be realized through a unique reference -*the TLM platform*-. Indeed, TLM platforms provide adequate and accurate hardware information for software designers much earlier than the conventional platforms such as RTL platforms. This information must be sufficiently accurate for software designers to start developing, testing, and debugging the software code closely associated with the hardware *without* pointless delay following the initial software development. In parallel, hardware designers can develop RTL platforms aimed at timing-accurate simulations, which are eventually employed for logic synthesis.

By the time the RTL design is complete, the software will have already been thoroughly verified on TLM platforms. The software design is thus ready to be integrated with the RTL hardware platform for system validation within a much shorter time than the traditional approach. As a result, sound and solid concurrent engineering is achieved through the unique reference of TLM platform.

A closer study clearly reveals that software running on TLM platforms can be classified into different categories according to their relationships with the hardware platform. This chapter will discuss extensively on the software categories ranging from design requirements to the mutual expectation of benefits between software and its hardware counterpart. Lastly, the chapter will draw a conclusion on how close collaboration between hardware and software developers could lead to a virtuous circle.

2.   SOFTWARE TARGETED FOR TLM PLATFORM

Throughout the development of a new SoC platform, various teams participating in the hardware design are always interested in running software programs on the platform. Be it any team varying from RTL design to functional verification and integration, early software execution means early catching of hardware or software problems. More essentially, executing software on the target platform helps to identify any potential mismatch between software and hardware designs.

In spite of its very attractive advantages, getting ready the software for early phases of SoC design cycle should never be done at any inappropriate cost of software development. The software should be executed on a development platform that is as close as possible to the final hardware platform. That will increase the probability of software reuse on the target platform with very little or virtually no modification on the subsequent hardware platforms. Such reuses trim down not only the overall software development time, but also the cost of refining software for these platforms.

A key parameter of developing the software targeted at running on TLM platforms is the immediate usability of the software in the current hardware design process. It is not quite convincing to claim a software piece being developed early in a project *useful* if that software piece could only be validated on a later hardware platform. The software must be tested on the target hardware platform while it is being developed. To bring the software and hardware design in parallel, they must be managed in tandem for scheduling smooth meeting points that optimize their mutual enhancements.

Running software programs on TLM platforms may appear easier than what it could really be for several reasons listed below:

1. TLM platforms are *not* real hardware platforms but abstract models for new platforms or IPs under design. To reach optimal uses of TLM platforms, software adaptations might be necessary.
2. TLM platforms have diverse modeling varieties. Each model might involve subtle adjustments in the software to adapt for non-fully covered features such as interrupt request (IRQ) or input/output (I/O).
3. Software compilations might require specific coding rules for proper program-runs in certain simulated environment of TLM platforms, for instance, compilations for handling timing issues on inexactly timed platforms.

All these reasons seem coercive on the software development using TLM. These good reasons, however, will definitely lead to efficient software coding and better code quality if they are appropriately practiced.

2.1   Adequacy of Software and TLM Platform

2.1.1   TLM Platform Accuracy and Availability for Software

The software development through the TLM approach depends closely on the modeling level of the corresponding TLM platform, which directly reflects the level of accuracy of the target hardware platform.

TLM platforms not reaching a minimal level of the functional behavior of the real platform may mislead designers to an erroneous software development by masking certain mistakes or bugs. The harmful consequence would be giving the wrong impression that the software is validated and ready to run on the real hardware platform. If a TLM platform poorly simulates the final hardware, very few software programs will be able to run correctly on it. It may miss testing critical features for hardware validation. The amount of time spent in such software development will be wasted and hence a higher global time-to-market.

On the contrary, it is sometimes unnecessary to have all design features simulated in TLM platforms if the whole process of concurrent hardware/software engineering is not significantly improved. Consider the following situation: Running natively compiled software on a timing-accurate TLM platform will *not* give any clue to the final software performance on the target platform. For such case, instead of developing timing-accurate TLM platforms, it could be easier to insert annotations obtained from cross-compilation into natively compiled software codes for studying software performance. Such annotations provide accurate statistical timing information without considering hardware features like cache, memory management unit (MMU) or write buffer, which could heavily influence the software performance in simulation.

Executing a software program on various functional TLM platforms has resulted remarkable outcomes. As an example, running a JPEG decoding program either on a PentiumIV with 1Mbyte of internal cache or on an ARM926EJ-S with 16Kbyte of internal cache may yield vastly different performance results of latency and throughput. The results of executing the software on TLM platforms help to better analyze various aspects of the hardware and software relationships. The software efficiency and correctness on the simulated hardware or hardware modifications for facilitating software development are examples of such potential improvements.

More importantly, running software on functional TLM platforms brings mutual benefits to the two working parties:

- *Hardware Developers*

  A live picture of how software programs utilize TLM hardware interfaces for real applications, which subsequently helps to improve the functional view of IPs on hardware platforms.

- *Software Developers*

  A live picture of how TLM hardware IPs react when software programs are executed on TLM platforms, which subsequently helps to improve the software implementation.

Indeed, these mutual benefits require not only the appropriate modeling choices of TLM platforms tailored for varied software design purposes, but also the proper manner of developing software in the right perspective of TLM platforms available at different design phases. Such careful matching of software development with TLM platforms is what we mean by the "adequacy of software and TLM platforms", which aims at optimizing the software development through the TLM approach.

2.1.2   Layering Software in TLM Platforms

To achieve such optimization, the software should be developed in progressive layers corresponding to the different levels provided by TLM platforms for simulation. This idea is illustrated by the development of a software driver for a UART sending and receiving characters on a given platform. The coding approach normally begins with a character-by-character interface, although a direct memory access (DMA) can be used on the final target platform. In the early design phase, an added-value feature like DMA may not be available yet in the hardware platform; besides, adding DMA in the TLM platform may cause some undesirable time delay in simulation. Most of all, it might be inefficient to use DMA for handling just a few characters because more management of registers and more software managing I/O blocks will be involved for the same number of interrupts. Thus, it is best at this point to start developing the driver without supporting DMA.

The good practice of the "layered" software coding through the TLM approach is strongly recommended. This concept is illustrated in the example of splitting a UART driver development into five phases as described in Figure 4-1. In the figure, each phase is represented by a task box. The size of each task box reflects roughly the relative amount of work dedicated for that particular phase with respect to the overall development.

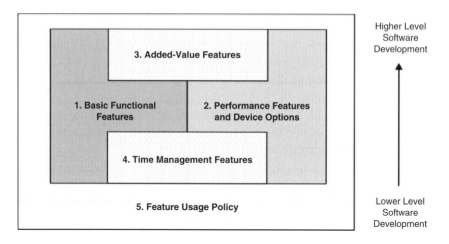

Figure 4-1. Layered Software Development

1. *Development of basic functional features.*
   In this example, first phase focuses on developing a functional UART driver managing simply character-by-character I/O interface. It is fast to be developed for an early interface testing.

2. *Development of performance features and device options.*
   Second phase develops performance features and options of the UART driver such as DMA access and cache management. Usually, these features can be easily inserted within the static conditional compilation.

3. *Development of added-value features.*
   To build a complete functional UART driver, all added-value features are developed in third phase; for instance, sleep/wake-up mode or performance counters. These features may be essential to help software designers in developing application software at higher level.

4. *Development of time management features.*
   Fourth phase concentrates on developing time management features of UART driver such as those for sleep mode or I/O completion delayed interrupts, which are dynamically configurable through external parameters. These features are typically very close to hardware view.

5. *Development of feature usage policy.*
   The final phase of "layered" software development determines the policy of how and when all the optional and performance features should be strategically employed. The mechanism of using all the features developed in the four phases earlier is carefully refined in this phase.

From the performance's point of view, the cost of TLM transactions is not very dependent on the amount of the data transmitted. In the example of the UART driver development, sending the entire text of a message using a DMA will be much faster than using a character-by-character I/O. Although the layered approach is valuable for the software development, using character-by-character I/Os in TLM platforms is inefficient due to their very long testing time: Assume that a character I/O takes N register accesses in the UART IP, i.e. each character will require N TLM transactions. Suppose that for every UART access, the DMA makes M register accesses. If the DMA is enabled, each DMA access to the UART IP can include any number of character I/Os, which will require only M TLM transactions. This will certainly utilize TLM platforms much more efficiently. Therefore, such performance features should be considered early enough in the design cycle to increase TLM platform overall efficiency.

To conclude, there are three rules to respect for the optimal software development and execution on TLM platforms:

1. *Do not develop software too much in advance*. It is not worth developing the software for hardware features available very late or prone to change in the future. Time saving may turn out to be worthless or extra delay may occur when hardware pieces are available or modified later because of the adaptation time.

2. *Organize software development tightly coupled with hardware design in layers adapted to IP functions*. Basic but complete features should be clearly separated from optional parts. These features should be incrementally tested in phase with their addition in TLM platforms.

3. *Give priority in developing performance features and device options for better software performance on TLM platforms*. If this is not appropriately done, software developers may not make the most efficient use of TLM platforms (they may probably get discouraged to use the TLM platform due to its slowness).

2.2 Analyzing Software on TLM Platform

As presented in Chapter 2, TLM methodology offers two distinctive models of the hardware platform for software development, namely untimed TLM (PV) and timed TLM (PVT). The current section focuses on how the software should be adapted for running on different models of TLM platforms.

Practical software properties will be provided throughout this section to demonstrate the global software quality improvement that could be brought by each TLM model. Such improvements will be compared to what RTL models and real chips can do for the software development today.

2.2.1   Functional Accuracy

TLM platforms are designed to provide an accurate *functional* view of the final hardware platform so that any software with correct *functional* behavior will be able to run on them. This may not include running the software with some non-functional aspects of the hardware platform such as real speed, linear time, or event ordering.

On top of the layered software development explained earlier, writing the software that is independent of any timing or event ordering issues is another good coding practice reinforced by TLM. For example, assume that an I/O starts with a register-write. The associated software should be ready to receive the I/O completion event at any time starting from the return of register-write operation. The same sort of the software functional behavior can sometimes occur on real chips because of I/O errors or suspended instructions due to interrupt handling. An untimed TLM platform, however, can offer the same advantage at much earlier availability!

Another example of analyzing the software functional behavior is described hereafter. Imagine that a software code reads some data from an always-ready source and writes it to a sink. In the real life, the sink will take some time to handle the data before it is ready to consume more data. Meanwhile, that extra delay will allow the software to perform other tasks. In untimed TLM platforms, the sink may accomplish the task instantly or in very small simulation time. The software will thus be ready to keep getting data from the source and passing it to the sink. If the software is not able to handle such behavior, it will spend all its time moving data from the source to the sink but nothing else! Running such software on TLM platforms will give the wrong impression that the functional behavior of either hardware or software is incorrect.

As long as the TLM platform is functionally correct, it will provide an absolute time reference with strict event ordering although it may not be time-accurate. Indeed, the root of the problem above is writing the software with the assumption that the sink will take enough time to handle its data to allow other tasks being scheduled. Two methods can handle this situation properly:

1. Let the software manage its tasks in the round-robin such as simple executive runtime.
2. Let the software handle the I/O management on an event basis. It will require some software adaptations for TLM platforms. The same problem may still occur if the software has too many events to

manage. This method, however, helps to handle certain rare real life situations that are probably never really tested in real chips.

This example clearly illustrates how the software should be adapted for the chosen model of TLM platforms for an appropriate analysis of the software functional accuracy.

2.2.2   Global Time Accuracy

The global time accuracy of TLM platforms is not an easy aspect to handle. The reason is that a system should be able to run even if it is *not* time-accurate. Since timing is very often an important feature for the software, untimed TLM platforms cannot completely ignore the timing behavior. Instead of implementing the full timing, events are strongly ordered within each IP. There is *no* global order for events occurring in different IPs, meaning that delays between event occurrences of different IPs are not accurate.

Implementing the global time accuracy in the software is not particularly difficult. The software, however, must be ready to manage this behavior proficiently. It is a bad coding practice to assume the order of two event occurrences in a system. For example, a timeout should be programmed to occur anytime after its scheduling without assuming that it may not occur before something else.

The major difficulty of implementing the global time accuracy in the software is the task management based on timing but not on event, for instance, time-sliced scheduling of Operating Systems. Such implementation is usable only if the software can ensure that a task is able to complete a sufficient amount of work before a time-slice. The system could otherwise be reduced to switch from task to task with little or no time to perform anything useful in between! In this case, the software may appear functionally correct but the execution result could be too far from the expectation of software developers. Software cannot do much to solve this problem. Rather, the hardware platform should give some hints on the time evolution such as estimates of software time expenses. When running software on untimed TLM platforms, software developers should somehow be ready to see some unexpected timing behavior of their programs.

In contrast, it is quite a different matter to handle the global time accuracy of the software on *timed* TLM platforms. Such platforms are able to provide the global time accuracy, i.e. a strong ordering of events for the entire platform. The software can thus be executed more accurately with respect to its timing behavior, including timeout, time-slice, or delay required by platform IPs in handling I/O events. Unavoidably, such timing accuracy is paid by a much less efficient software execution because there are more events to manage compared to those in untimed TLM platforms.

The global time accuracy of a given platform depends very much on the way of how IPs are implemented in the platform. If all IPs comply with the timed TLM constraints, the entire platform will be globally time-accurate. Software programs may run only with approximate timings on the hardware platform in cases where certain IPs are not timed TLM compliant, or native software compilation or non time-accurate ISS is employed. Nevertheless, it could be interesting to test the software in environments that are different from the final timed platform.

Obviously, it is more understandable to develop and test the software on timed TLM platforms with fine-grain timings than on untimed TLM platforms with approximate timings. The most suitable choice for analyzing the software behavior related to the global time accuracy is of course the timed TLM platform. Software programs, however, should run correctly without any code modifications on both untimed and timed TLM platforms.

2.2.3   Protocol-Timing Correctness

When an external component is connected to a SoC, software developers need to program the relative timings correctly for eliminating any potential communication hazards. This is probably one of the trickiest problems to solve in the software because its failure cannot be easily detected on RTL hardware platforms. The symptom of such problem is typically an unstable system that works properly for some time, but crashes suddenly with no warning signs.

Timed TLM platforms are the best spot to uncover such programming errors. For example, PVT platforms can effortlessly reveal insufficient wait states for accessing a memory IP by comparing the number of wait states programmed by the software to its internal characteristics. To do so, the PVT memory controller validates if the time amount required by the memory access is coherent with the number of wait states programmed. If the wait states are insufficient, the memory IP can send a notice thanks to the timing information held by TLM transactions.

The concept explained in the example above, by analogy, applies to any other external controllers connected to SoC platforms via standard industrial buses such as I^2C, CAN, I^2S, SPI, and so on. Once the first prototype board around a SoC platform is built, it is usually too late to fix an external protocol-timing problem where platform controllers and external devices sharing the same protocol fail to communicate. A "quick and dirty" way to overcome such hardware problems is to modify the software, which unfortunately results in, most of the time, reduced performances and functionalities.

Protocol collision management is another protocol-timing test that can easily be set up thanks to TLM platforms. Some simple bus protocols such as CAN or I²C are designed to solve collision issues by forcing a master to be a slave, which will consequently change the behavior expected by the software. Protocol collision is a very difficult software behavior to test because forcing collision on hardware is a tough procedure that usually requires special hardware to test all potential cases. Although a bus-cycle accurate platform can set up all types of collisions, timed TLM platforms are sufficient to set up global collision required by software developers at an earlier availability. In addition, the input of the TLM platform could be programmed to show such specific hardware behavior. Thus, it can provide software developers with the ability to validate the actual software behavior on demand.

2.2.4 Resource Overflow

With the advent of SoC, software developers have somewhat lost a little of the control they used to have over the unexpected limit reached by performance. Consider the following case of resource overflow: a 100Mbps Ethernet controller together with a fast CPU can sustain an Ethernet flow close to the theoretical limit, particularly for full duplex mode without collision on wire. If the theoretical limit is far from being reached, software developers can use a packet analyzer to examine the packets received by Ethernet driver from the controller. They might sadly notice that the packet is surprisingly in coherence with the speed announced by the application. The only solution is to analyze deeper the packet flow between its input in the Ethernet controller, and the interruption signaling for its availability in the memory.

In general, it is extremely hard to peek at the activities going on inside a SoC. But, there are so many hardware items involved in the packet management (IP, DMA, buses, caches, etc) that it is almost impossible to easily detect any bandwidth bottleneck. Resource overflow, on top of this difficulty, is very often hidden by some hardware limitations in bandwidth, access priority, etc. All these factors make this specific problem a real tough job to fix for software developers. In addition, RTL platforms are not exactly the right solution due to their performance limitation.

A good tactic to cope with resource overflow will be employing TLM platforms because they provide adequate details and hints to guide software developers in locating the problem. Timed TLM platforms optimize timing measurements to avoid all hardware contentions in accessing resources on the platform. As a result, it is easier to get the best performance measurements especially for cases where cycle-accurate ISS is applied. If the performance is satisfactory, software developers can proceed with a bus-cycle accurate platform, which gives results on the miscellaneous hardware contentions that the system has to face for this particular test. With all these results, software developers will be able to locate the problem of resource overflow.

2.2.5   Performance Profiling

The foremost interest of executing software programs on TLM platforms is of course getting the software running on the target platform. Once the software gets up running properly, the next goal will be collecting early performance results before the final hardware is available. Performance measurements are not only based on timings, but also start with non-timing counters such as the volume of transactions exchanged by IPs. This job can be accomplished adequately by untimed TLM platforms.

Untimed TLM platforms, however, cannot do much to obtain timing results. Attempting this on hardware platforms may not be the best choice because the measurement software itself could modify the overall timing of the platform. Since measurement mechanisms are embedded in the IPs, timed TLM and RTL platforms are both capable of evaluating timing results without altering the overall timing of a given platform. Obviously, timed TLM platforms are better options than RTL platforms for performance profiling thanks to their usual earlier availability.

Inconveniences may arise in common practices of performance profiling. Frequently, software needs to be modified to obtain profiling results. The measurement software is thus intrusive on the system platform. Sometimes, the profiling procedure could be dreadfully time-consuming or the external hardware required for extracting profiling results from a platform may not be available all the time.

Through timed TLM platforms, however, all these inconveniences are straightforwardly resolved. Since measurement mechanism is embedded in the platform IPs, performance profiling is *independent* of any software running on the platform. That will greatly reduce the workload of software developers.

The example of latency profiling gives a better idea of how helpful timed TLM platforms could be for software performance profiling. Latency is very hard to finely measure when the software is running on the hardware platform. Such difficulty is particularly bitter for real-time systems that are extra-sensitive to latency issues. Timed TLM platforms, nevertheless, can run real-time software without any modification to conduct profiling such as building the histogram of interrupt latency. Therefore, a software developer can get fine and accurate results without any modification of software, just by extracting the right profiling from its timed TLM platform.

2.2.6   Hardware Utilization

Running software on TLM platforms grants the ability to detect whether software makes the *right* use of hardware platforms. Additional non-functional code can be embedded in TLM platforms to validate if hardware is utilized properly as expected by its design. Although hardware could tolerate certain bad or poor utilization by software, the resulting effects of such use are sometimes likely out of software expectations.

Consider the example of UART transmit-character register. Under normal practices, it is not permissible to push another character into this register if the previous character is not yet consumed. The hardware, however, allows software to freely write characters in this register as many times as it wants, without any effect on the IP behavior. Most of the time, overwriting character in such manner is a programming error. TLM platforms can help to verify the same sort of programming errors without much effort. As a result, software developers can obtain reliable hints on the potential programming errors in the software.

TLM platforms also provide interesting results about the software utilization of particular hardware features. For the same register in the last example, certain UART IPs allow software to push another character in the register while the current one is being transmitted. This is a special feature to reduce the latency between the end-of-transmit interrupt and the availability of the next character to be transmitted, which software developers are invited to use as much as possible. Internal counters can easily be enabled to measure how frequently this hardware feature is used by the software. Following the simulation, a statistical listing for the utilization of special hardware features can be provided. Based on the list, software developers can learn better about the hardware utilization by their software implementations, whereas hardware developers can see the actual utilization of hardware features in real cases.

2.2.7   Conclusion

After discussing on how untimed and timed TLM platforms can help software developers, Table 4-1 summarizes and compares the different kinds of software behavior that can be studied at different modeling levels.

At first glance, the summary may mislead to the conclusion that bus cycle-accurate (BCA) platforms give the best software support. This could probably be true if the overall platform performance and setup work are *not* considered. This is the reason why these two criteria usually determine the interest level of using a TLM platform model for running, testing, and debugging software before RTL and real hardware platforms are available.

If these criteria are considered, BCA is certainly not the best option because both untimed and timed TLM still provide faster performance than BCA, and are usually set up and integrated much quicker. Although RTL is the slowest for performance and construction, its vital hardware simulation capabilities make it necessary to be constructed (normally after TLM platforms). Concisely, TLM platforms are the most compelling models for running and testing software before the real chip is available on silicon wafer.

Table 4-1. Software Behavior Observed at Different Modeling Levels

Software Behavior	PV	PVT	BCA	RTL	Silicon
Functional Accuracy	Yes	Yes	Yes	Yes	Yes
Global Time Accuracy	No	Yes	Yes	Yes	Yes
Protocol-Timing Correctness	No	Yes	Yes	No	No
Resource Overflow	No	Yes	Yes	Yes	No
Performance Profiling	Yes/No	Yes	Yes	Yes	Yes/No
Hardware Utilization	Yes	Yes	Yes	No	No
Accurate Concurrency	No	No	Yes	Yes	Yes

PV = Untimed TLM   BCA= Bus-Cycle Accurate
PVT = Timed TLM   RTL= Register Transfer Level

Notice that the accurate concurrency is a behavior listed in Table 4-1 without being discussed earlier. This is a critical behavior to analyze when two or more IPs try to access concurrently the same platform resource like bus or DMA. Such concurrency is part of the functional accuracy that can be implemented in TLM platforms. The accurateness of such concurrent collision, however, is not handled by TLM because it requires cycle accuracy to manage the interactions and requests of platform IPs.

2.3   Software Environments of TLM Platform

Running software on TLM platforms depends not only on the platform design, but also on the different environments in which the software will be handled. There are four major TLM software environments, which will be discussed in the coming sections:

- *Software Development Environment*
  Describe how software is produced and debugged.

- *Software Execution Environment*
  Describe how software is executed on TLM platforms.

- *Software Integration Environment*
  Describe how software is integrated into TLM platforms.

- *Software Simulation Environment*
  Describe how software gets input data and puts output data.

As depicted in Figure 4-2, these four software environments correspond very well to the famous V-diagram for the life cycle of software. Each of the environments prepares the necessary setting for performing the different software work at various phases.

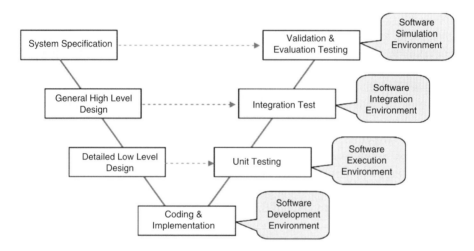

Figure 4-2. Relating TLM Software Environments in V-Diagram

2.3.1 Software Development Environment

TLM offers the great advantage of having a simulated hardware platform that can be either natively compiled for faster speed or cross-compiled for binary compatibility and higher accuracy. This dual compilation capability therefore provides two development environments to software coding and implementation.

The *cross-compilation* development environment requires embedding a model for the targeted processor (usually called an ISS) in the TLM platform. The software can then be compiled for the actual target processor and simulated by the ISS. The software is thus isolated from the TLM platform execution by the host system.

The *native-compilation* development environment merges the execution of the software with the execution of TLM platform IPs by the host system. The software is link-edited with the TLM platform simulation code and executed as part of the complete platform process; the main characteristic is that software shares the same address space as the platform simulation code itself.

TLM software development environment relies heavily on the decision made for software integration. Different integration methods require different integration tools, for instance, native integration necessitates different tools from cross-integration. According to the opted integration method, the appropriate development tools must be applied; and that will determine the software development environment.

Certain development tools, however, remain the same for either native or cross integration. A handful of examples include editors, source code generators, and particular compilation suites such as those using GNU tools. Sometimes, it is even *compulsory* to keep the same tools. Consider the example of GNU tools: If GCC and binutils are used, source code must be compiled exactly the same manner in either native or cross environment. The reason is that different compilers may actually require code adaptations due to their different specific syntax or extensions.

Using two different development environments (and thus two different integration environments) reinforces software portability, especially if both have different compilers. Not only can the code quality be improved by porting the software on two distinct environments, but more potentials problems can also be uncovered through different code compilations.

Conversely, software may undergo the side effect of being sensitive to certain processor aspects listed below due to using two different central processing units (CPU) in its development environment:

1. *Endianness*. The software must be ready to support any endianness (little, big, reverse, cross, etc) if the two processors (native and cross) have different ones.
2. *Assembler*. If the software embeds assembly codes as C extension, the same function ought to be available for both processors; it should otherwise be replaced by a functionally equivalent but less performing C code.
3. *Self-modifying code*. If the embedded software uses self-modification, a similar feature must be made available in the native environment.
4. *Data alignment and size*. If the embedded software relies on specific data alignment and size, then the software must provide all used compilers with these requirements.

5. *Addressing features.* If the software relies on specific addressing features imposed by the final processor, they must be implemented by any potential native processor.

In the software development chain, post-compilation tools for debugging and profiling could be very different between native and cross-compilation. Software debuggers, in particular, can be totally different. The native debugger controls the running platform directly whereas the cross debugger controls the platform indirectly via a client-server architecture. Showing too many tiny details of the TLM platform to software developers is an additional problem of the debugger in native environment. It could be very confusing for those developers who wish to debug their software but not the hardware. The native debugger should then be adapted to display only necessary information to software developers.

Compared to debugging, software profiling on TLM platform is quite a different matter. It is only worthwhile for special cases as follows:

1. *Profiling conducted on natively compiled TLM platforms.* Although the results can be very different from the final platform, it gives some valuable hints on the behavioral performance of the platform during early development phase, such as access counters. Calling graphs might also be extracted in such profiling for early performance and execution path analysis.

2. *Profiling conducted on cross-compiled timed TLM platforms.* Such profiling provides the very first idea of software profiling with coarse-grain timing before RTL hardware platform is available.

2.3.2   Software Execution Environment

TLM software execution environment is determined according to the adopted development environment. Software reaching this phase should be ready to be executed for unit testing, either as native compilation or as cross-compilation with an ISS.

With such performance-reducing factor as ISS overhead in cross-execution or hardware emulation, native execution is certainly the fastest execution environment. This is nonetheless not always a true statement because timing issues in natively compiled codes are totally different from those in cross-compiled codes. For an example, a different timing in hardware could probably cause such an overhead that the software is paralyzed, or it could probably run the software correctly in the native mode but masking some stubborn bugs that would only be visible under ISS execution!

One of the assumptions held in the previous example is that the compiler chain produces correct code in both native and cross cases. Running native codes can help nothing in debugging cross-assembled parts or coprocessor specialized instructions. In addition, certain data representations cannot be compiled because they are unavailable on native platforms, for instance, floating-point representations.

The toughest challenge in software execution is the memory mapping of the software. It is quite straightforward for software cross executed with an ISS. The software simply runs in the memory space defined by the ISS, i.e. the memory zone perceived by the software in the platform. The situation, however, becomes trickier in native execution. The software is bound to run in the memory space defined by the local host, which could be different from the one programmed in the software for the final hardware platform.

Consequently, the software needs to be relocated into this different memory zone. Addresses of memory layout might need to be translated to addresses not used by the underlying host system. Some software adaptations are required to remap cross-compiled hardware addresses into natively-compiled addresses without flaw. There is something similar to implement for register accesses. The reason is that they are not simple memory-mapped read and/or write accesses as in the cross-compiled environment, but requiring some modifications to fit the actual bus modeling schema.

The register access remap should never be regarded as useless overhead, but rather as a good software coding practice. It allows the re-definition of hardware register accesses via generic read and/or write macros according to different compilation modes. Native compilation paves the way for software developers towards the first functional view on the final hardware platform; meanwhile, it enables the implementation of valuable portability features in software.

Among all the possible native execution environments, the operating system (OS) emulation deserves a special hat's off. Its goal is to abstract the interface between the OS and the hardware platform to set up a native environment. In this environment, applications can run natively on the OS layer while the OS itself can run natively as well on the hardware platform. Since the CPU used in the host machine is more powerful than the one in the real hardware, such setting can reach very high performance by running software on the simulated platform much faster than on the real hardware platform.

2.3.3 Software Integration Environment

TLM software integration environment provides the right setting to perform integration tests for a given system. It is not a simple task to determine how TLM software should be integrated into a hardware platform, especially when multiple solutions exist. One of the solutions is to incorporate the embedded software into the simulated hardware. It suggests that the software interacts with the hardware in terms of reading or writing data. These interactions are simulated as software actions on TLM platforms. For example, a hardware IP register access is interpreted as calling the right function in the IP module of TLM platform to simulate the access.

When modifications are necessary, it is preferable to change the software instead of the hardware for the reasons of cost, time, and workload. Therefore, it is sometimes desirable to separate software from hardware. An alternative solution could be compiling software for the target CPU and simulating IP accesses through an ISS.

Bear in mind that performance is one of the main criteria for using TLM platforms. In the alternative solution, performance is yet a problem because ISS is not as fast as native CPU. If performance is the main consideration, the most appealing solution could be native execution. The software must then be link-editable in TLM platforms, and that could probably be a source of diverse problems. Some of the possible problems are listed below:

1. TLM platform is link-editable through some external libraries that must be compatible with those of the software. If they use different or incompatible versions for the same library, the integration will fail because the same symbol may cover different functions.
2. If the software defines external symbols that collide with those of TLM platform libraries, the same problem as in (1) will occur.
3. The software is obliged to compile with the definitions of TLM platform that could potentially collide with those of the software.
4. The software may use process resources such as signals, memory mapping, and file descriptors in an incompatible manner with those on TLM platforms.

This list is non-exhaustive but enough to show the lurking problems that could appear anytime during the integration process. It is therefore hard to decide beforehand if native execution is feasible, although it may appear attractive in performance. Anyway, a potential solution always exists, i.e. integrating software into a cross-compiled environment where the software runs independently of its hardware platform.

2.3.4   Software Simulation Environment

Once it has managed to execute and integrate correctly on the target platform, TLM software will proceed to the software simulation for validation and evaluation testing. The software usually cannot run alone in the simulation environment because the entire board holding the SoC is involved; meaning that some external input and output data flows are required to conduct such simulation in a real environment.

A simple way to establish connections between the platform and the external world is to input/output data of platform IPs from/to local host files. Its greatest advantage is the easy setup that enables software to run test samples promptly from the local host files. Such reference samples will really be handy for debugging algorithm or platform behavior of certain final code, say protocol decoding.

Connecting with local host file is not always sufficient. It is interesting to connect IPs with real devices in certain cases; for instance, interfacing a card reader IP with a serial line, or bringing the actual character protocol into an UART IP to allow testing software on emulated hardware that is connected to real hardware. Such "real" connection can also be employed for buses like Ethernet or USB through the host system devices.

Another interesting aspect of the TLM simulation environment is its ability to report the input/output of hardware multi-media to the host. Consider the following example. If the software is designed to use an LCD of a given size, it is quite straightforward to map the LCD on the host graphical window. That allows debugging the exact contents provided to users without needing to write a single line of code, which is anyway not reusable on the final hardware platform. Essentially, this aspect is the most remarkable difference of TLM simulator from an emulator that really entails interfacing with the software.

The greatest interest of exporting the simulation environment out of the platform is to provide total flexibility in the way of connecting the platform to the external world. Defining standard interfaces for internal IPs is a corollary of giving software developers such flexibility in the simulation. With this flexibility, software developers can simulate their design with the external world in any way they wish (including incompatible simulations), and to any extent they wish (up to the complete simulation). The platform with such interfaces needs not to embed any external input/output devices such as graphical windows to simulate serial communication. As a result, the platform is more portable from one system to another because the communication will be standardized via an open socket protocol.

2.4   Conclusion

TLM platforms provide software developers with a brand-new interesting methodology to test the software in a hardware simulation environment. Since the simulation is pure software, it is possible to set up different environments depending on the characteristics required by the platform, including accuracy, performance, connection to the external world, and so forth. In fact, TLM has filled up the gap between software and hardware developers. A bridge is now constructed between these two teams to enable each of them to observe from their own perspective how their development work is used by another team.

2.4.1   TLM Impact on Software Development

TLM platforms provide software developers with a hardware base to develop and more importantly, to test their software long before any pure hardware emulation is available. This is particularly helpful for the new hardware IPs on which no software has ever been ported or written yet. The major advantage of such early software development and testing in the SoC design cycle is to reveal any potential problem between hardware and software prior to their delivery.

Developing software that can be simulated immediately on the target platform is certainly beneficial. It helps to produce better software implementations in terms of portability and hardware utilization. In general, TLM reinforces good practices in software development process.

Based on TLM platforms, software developers can fully focus on the coding targeted for the final hardware platform without building any temporary dummy (and sometimes costly) hardware platforms. The software can be simulated at different accuracy levels on TLM platforms in the different environments required by the software developers. Such conveniences grant software designers ample freedom to perform their job without waiting keenly for the first hardware platform.

2.4.2   TLM Impact on SoC Design Flow

The overall SoC design flow has to be reconsidered when using TLM. This is essentially the foremost impact of TLM on the SoC development. A TLM platform is regarded as the first hardware prototype wherein software developers can execute their code. Even a partially complete TLM platform can interest software developers because it can already help debugging their code up to a certain extent.

In brief, TLM can significantly alter the conventional manner of how a system-on-chip is constructed by creating more positive interactions between hardware and software fellows. A veritable hardware/software co-design will therefore be achieved through TLM approach.

Another appealing advantage of TLM is the cost. The number of a given TLM platform can be multiplied as many as the host machines that it can use for running. Consequently, the number of software developers being able to use this particular TLM platform is potentially unlimited at a given time. This advantage can rarely be provided by a typical hardware prototype such as emulator due to the cost issues. Naturally, more engineers will be able to work on a SoC project in its early design phase based on TLM platforms.

Figure 4-3 illustrates the time phases of the TLM-oriented hardware and software development. During the development of the untimed TLM hardware platform, a huge functional part of software programs can be developed. Once the untimed platform is ready, software designers can start testing the written software on this platform. Certain time-level features can be added to the software codes based on the untimed platform. Through observing the software execution on the untimed platform, we can improve not only the codes but also the untimed platform. Meanwhile, hardware designers continue their work in conceiving the timed TLM hardware platform. Once it is done, the further developed version of software codes will be executed and tested on the timed platform. Based on the timing information on the timed platform, software designers can further develop the software for the hard timing parts. Such software execution helps not only to improve the software code but also the timed hardware platform. At the same time, hardware designers keep on their job to conceive RTL hardware platform. Note that as the RTL hardware is ready, the software will have already been well tested on untimed and timed TLM platforms. Such "almost-final" software applications will be able to run quickly on the RTL platforms to reveal some hidden stubborn bugs.

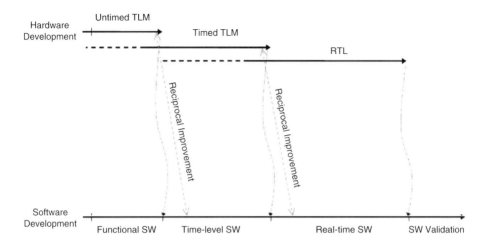

Figure 4-3. Time Phases of TLM-oriented HW/SW Development

The interactive design between hardware and software teams enhances the whole system design by visualizing their work to each other in a transparent manner. Hardware designers can observe how the software program utilizes the hardware platform while software designers can see how the hardware platform reacts to the software execution.

2.4.3 Illustration of Software on TLM Platforms

After reviewing various aspects of the relationship between software and TLM platforms, it is worth our time to discuss in details about the development of different software families based on TLM platforms in the rest of this chapter. The discussion will lay emphases on the objectives of using TLM platforms, TLM-based development and execution approaches along with illustrations of practical examples.

From an architectural point of view, software can be arbitrarily split into three layers as depicted in Figure 4-4. Each layer has a particular relationship with the hardware, and thus with TLM platforms.

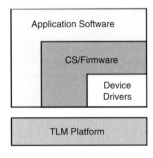

Figure 4-4. Software Families Developed on TLM Platform

3.   TLM-ORIENTED DEVICE DRIVERS

3.1   Introduction to Device Driver

Device driver is the closest software level to TLM platform as shown in Figure 4-4 earlier. The key role of device drivers is to abstract low-level peripheral details to represent a generic programmable interface comprising a number of predefined functions. Device drivers should be the only entity accessing peripheral resources such as registers or shared memory.

A common method of accessing peripherals is via register accesses. Usually, registers are gathered into a unique I/O memory area reserved for specific IP accesses. Since their behavior is peripheral-dependent, register accesses must be correctly implemented in TLM platforms with the accurate functions. Another way of accessing peripherals is by means of shared memory. A memory zone is reserved in TLM platforms for data exchanges between peripherals and device drivers. Such data exchanges are performed within the structures defined by the peripherals.

3.2   Purposes of TLM in Device Driver Development

3.2.1   Unit Test Development

One of the very fundamental purposes of device drivers is to develop unit tests for a given IP on a TLM platform. Such device drivers run simple tests to assure the proper implementation of platform IPs. The degree of correctness tested by them depends on the types of the underlying TLM platforms, for instance, it is out of scope to test timing issues of an IP on the untimed TLM platform.

A device driver may cover more than a single IP if a DMA is coupled with the IP-under-test. In that case, the DMA will be tested as well but only for its interactions with that particular IP-under-test, i.e. the device driver can only conduct partial DMA testing.

3.2.2   Non-Regression Test Development

Device drivers can be developed as simple software for performing non-regression tests on TLM platforms. In the early phases of TLM IP development, it is vital to run device drivers on the TLM models to verify their correctness. As the design develops gradually into TLM models and becomes more complex, running the existing device drivers can be considered as a good non-regression test suite, which can verify that the additional new features work properly without distorting the old features.

Non-regression tests are usually totally independent of whether there is an embedded processor or not within the platform. Thus, the same tests are portable on the different platforms integrating the same IPs. This is a great advantage to validate quickly the reutilization of IPs on various platforms. The only characteristics to modify from one platform to another will be the base I/O address and the interrupt mapping.

3.2.3   OS/Firmware Device Driver Development

The term "device driver" is indeed derived from the semantics of OS/Firmware. It represents a piece of software developed specifically to be inserted into another piece of software that is more complex, i.e. the OS/Firmware itself. The purpose of this extra software piece is to isolate low-level management of IPs in an independent module with some externalized interface.

Device drivers serving for such purpose do not run alone as in the two previous cases, but rather in an environment with some constraints that will impose a particular way to use IPs. These constraints enforce a conventional manner of software coding, which may potentially improve the way that hardware is programmed.

Despite some attempts to standardize the interfaces, device drivers are usually not portable from one OS/Firmware to another; hence leading to different ways of using a given hardware.

3.2.4   Experimentation of New Hardware Features

Another interesting purpose of device drivers is to exercise new hardware features for experimenting their different aspects such as programming ease, performance improvement, programming examples, etc. Such experiments can be quickly set up on TLM platforms to test tiny modifications on the hardware before the real alterations.

Since device drivers are final software pieces of larger models like OS/Firmware, they can be modified independently from the rest of the whole system to include new hardware features. It is therefore very easy to rapidly set up a model for hardware developers to build their intended design, and subsequently exercise this new design under a realistic software execution.

As a result, hardware developers are able to verify the correctness as well as the resulting effects of their tentative design under the real scenario of software run.

3.3 Approach to Device Driver Development

This section focuses on the different approaches to developing device drivers. General rules of writing software targeted for TLM platforms are presented in section 2.1. For device driver software, the methodology of layered software development remains valid for its development and testing. Some additional aspects that deserve special attention will be explained extensively in this section.

3.3.1 Interrupt vs Polling Management

Reporting occurrences of interrupt events within an IP is normally managed by setting a particular bit of the IP status registers. Optionally, the IP may forward a signal to an interrupt controller that will in turn monitor the CPU interrupt line. From the angle of software, there are two methods of managing IP events:

1. *Synchronous Programming.* Polling (i.e. reading continuously) the bit reserved for interrupt in the status register until the right value is obtained.
2. *Asynchronous Programming.* The software executes standard procedures. Under interrupt occurrences, it is diverted to execute a handler that has been previously associated to the interrupt. At the end of the handler, it will simply continue execution of the procedure at the point it has been interrupted.

The two methods are not really independent in TLM platforms. The first method issues a TLM transaction whenever the status register is read or accessed. It thus induces a lot of overhead especially if the interrupt event takes quite some simulation time to occur. This is the appropriate choice for coding interrupts in unit test software because there is normally no other software running than unit testing.

Waiting for an interrupt as described in the second method is very close to the real situation on the real platform. It leaves no impact on TLM platforms because the IP will initiate a transaction when the real interrupt is routed to the interrupt controller. This method is suitable for testing the interrupt mechanism of a system. It is particularly useful for device drivers as they need to continue other tasks while waiting for the interrupt event to occur.

The software should take into account some unexpected behavior that could probably be induced by TLM platforms. One of the common examples is the approximate timing estimated by the untimed TLM platform, which delays certain event occurrences. Asynchronous interrupt programming assumes that the hardware will notify event occurrences with sufficient delay, which allows the software to perform some useful job while waiting in background for the interrupt. The consequence is that the software may not be able to do anything or even spend more time than expected in the interrupt handler if the interrupt occurs too quickly. The same problem, however, will not arise in the timed TLM platform since the timing is an absolute reference, i.e. events setting an interrupt will consume the required time amount before their occurrences.

Therefore, polling should be applied as much as possible in the untimed TLM platform instead of asynchronous interrupts. However, if the asynchronous interrupt modeling is required on untimed TLM platforms, interrupts must be expected to occur at any time. They can even occur in the same instruction of the I/O that starts an interrupt, which can rarely happen in the real life.

3.3.2   Time Management

Unlike interrupt controllers, certain IPs such as real-time clock, watchdog or timer deal directly with time management. The software written for such IPs must be aware of the time events like time-slicing, time-out or time-count for running on TLM platforms.

Timing is locally accurate on the untimed TLM platform. From the software point of view, events are locally ordered within a given IP. Consider the following example: a given IP with two timers programmed for sending interrupts at different dates will always send interrupt events in the well-coordinated order. Now, consider two distinct IPs with a timer in each. Even if the two timers are programmed in the same manner, IP events could occur in any order because both IPs are completely independent from each other with unspecified relative timing approximation.

Luckily, it is quite uncommon to depend on the relative timings between different IPs to run a software program correctly. For instance, time-out is usually implemented on top of a timer by the software so that it can be ordered continuously. Although the simulated time difference remains unpredictable, it brings no problem since the software usually relies on time order but not on time difference.

Testing scope is quite restrictive for timing aspects on the untimed TLM platforms as the timing accuracy is not really measurable. Low-level design is thus reduced to validating the functions of interrupt and status indicators. The timed TLM platform, however, offers larger capabilities in terms of timing testing.

Another important point on timing is time-slicing. When a dummy or buggy C program executes a "`for(;;) continue;`" sequence, it will keep looping forever. It is always possible to stop this loop by sending an interrupt, e.g. character typed or time-out, which can divert the execution from the loop and eventually stop the loop. Running such programs on an ISS is well handled by TLM platforms. The untimed TLM platform manages this program by advancing its timeline from time to time, even if the ISS does not require any I/O on the IPs. For example, the time progression can take place when the ISS runs an I/O access; the internal SystemC scheduler can then be called freely to move forward the timeline. For the timed TLM platform, the rule is much stricter since timing accuracy is required. The ISS must access the internal SystemC scheduler (even for nothing) in order to let other IPs running their codes at the right scheduled time.

The approach is totally different for natively compiled applications as they are integrated into the execution environment of TLM platforms. Bear in mind that TLM threads are *non-preemptive*. If any thread happens to loop, no other thread can preempt it from looping and the TLM simulation will just loop forever. To let other threads run, a special thread layer such as OS emulation can be of great help by simulating multiple OS threads within the same SystemC thread. An alternative solution is inserting some calls to the internal "`sc wait()`" function at the right locations. This function will essentially give a chance to the system to progress its simulation. Such situation is one of the very few circumstances where the software must cooperate directly with the TLM platform.

To conclude, software running on the TLM platform, especially when natively-compiled, must be capable of handling unpredictable time management.

3.3.3 Performance-Accelerating Hardware Features

It is a general comment that TLM platforms do not simulate hardware fast enough. Although this is always a personal perception, such moderated simulation speed might actually be very useful to detect some problems that may appear unobvious on fast-simulating platforms such as the real hardware platform.

As a matter of fact, the reduced simulation speed is frequently a "bug amplifier". A subtle bug occurring for a very short time period could probably be invisible during the simulation on the hardware platform. The same bug, however, may turn into a disaster in a TLM simulation and thus much easier to be detected and fixed.

Suppose that a driver for a slow-communication IP does not use a DMA correctly. The real hardware platform works so fast that it may conceal this problem on regular uses. The problem can only be revealed by an integration test where other IPs are involved to use the slow-communication IP intensively. The system will give an abnormal response time that serves as an indicator of such problem. A TLM simulation, on the other hand, shows the abnormal response time immediately because such problem will give the character-by-character output (1 character per transaction) instead of the message-by-message output (N characters per transaction) that should normally be provided by the DMA use. Since the overhead of a transaction is not negligible, a unit test is usually sufficient to uncover the poor programming of the DMA.

The similar problem can be encountered in cache programming. If the cache is badly used or unused, the number of accesses to the TLM memory will be unacceptably high. Software developers will consequently notice a bus overhead rapidly, and thus identify a cache-related bug.

Therefore, the moderated simulation speed on TLM platform provides users with an early detection of misused features. This is extremely helpful for revealing those directly related to the overall system performance but hidden in a small local area for a long time. Indeed, these are very tough features to detect because they appear functionally correct. It is the reason why some software programmers may not see the advantage of TLM "bug amplifier" right in the beginning. Once they get more acquainted with TLM, they will definitely find this characteristic rewarding.

3.3.4 Peripheral Error Management

Another critical piece of device driver software is the management of peripheral errors. In common practices, this software piece is only ranked as secondary level of importance because the priority is always given to programming the regular peripheral uses. Unfortunately, the quality of a low-level code like device driver is not in the regular working parts, but rather in the error management and recovery.

Through modifying specific values in the setting, debuggers are used to "set up" and reproduce an error to facilitate the analysis of a particular fault in details. This method, however, will get a little cumbersome when an error comes directly from hardware devices. Too many registers will have to be set up in debuggers for such bugs. Some manual intervention or script-writing in debuggers is even required for certain cases. Consequently, such errors become extremely difficult to regenerate or reproduce "correctly and accurately", for instance, in non-regression tests.

Let us consider the error management of the Ethernet controller. Under normal working conditions, the Ethernet driver is not in charge of any errors. For high system load, the driver must nonetheless face plenty of severe conditions such as input errors, buffer underflow, out-of-buffer, etc. Under these conditions, the driver may decide to reinitialize the Ethernet controller while a simple recovery procedure could be sufficient. This technique works most of the time but it may result in catastrophic performance consequences. For this reason, it cannot give good quality software although it functions correctly.

Such hardware-related error management is a real pain for software developers. It consumes much time in understanding and coding yet brings too little visible functionality to the software. Most of all, testing errors that practically never occur in a real system is too huge a challenge. Hardware developers do have hardware devices to reproduce specific errors easily. However, these devices may not be available for software developers. Even if particular hardware test sequences can be set up, they will not be suitable for software error management.

TLM platforms are sound solutions for handling peripheral error management in device drivers. Software developers can simply inject data from the external world into the platform IPs to reproduce specific IP hardware errors. This error injection helps to test the behavior of device drivers when the error actually appears. Since the error is managed by the software, error sequences can be produced in the IPs as many times as required for running the error testing at high level of confidence.

3.3.5   Native Compilation

Native compilation is the fastest TLM simulation system for software. Although irresistibly attractive, it must nevertheless be employed with meticulous care for a number of potential pitfalls. In particular, software codes must respect the underlying restrictions rooted in the fact that the software is link-editable with TLM platform codes, i.e. TLM platform codes will be embedded together with software for running.

The most obvious restriction is the non-exclusive use of shared resources. Software must never "monopolize" common resources shared with TLM platforms such as heap memory, signals, file descriptors, etc. For instance, the signal handlers from software codes should never replace but add onto those already existing in TLM platform codes; in the same sense, the allocation order of file descriptions should never be deduced from the one of the underlying OS algorithm.

 Software codes must never be based on libraries or software compilation tools that are incompatible with those required by the TLM platform codes. A simple example can be illustrated by GCC compiler. It is well known that the GCC-2.95 release is incompatible with the GCC-3.x release for C++ programs due to changing of name mangling algorithm. If a software program compiled with GCC-2.95 is link-edited with TLM platform codes compiled with GCC-3.1, the link-edit will fail indicating that a problem exists or worse, the link-edit will seemingly succeed but the execution will crash without any obvious reason.

In the same line of idea, another interesting point is dealing with threads. Threads used in a software program must be compatible with those used in TLM platform; besides, they must respect the reentrancy programming constraints of TLM platforms. In other words, only threads compatible with SystemC runtime are allowed for TLM-oriented software codes because TLM platforms are based on SystemC runtime. For example, only one OS thread is permissible in the OSCI runtime, which restricts uses of SystemC threads and those simulated within a SystemC thread. In addition, the software thread scheduling has to be compatible with the one used in SystemC. The reason is that SystemC functions are not required to be implemented as reentrant; for instance, the current thread scheduling of OSCI runtime is neither reentrant nor thread-safe.

Debugging natively compiled software is much more complex as it is based on the SystemC runtime. There are two major difficulties. First, software developers may perceive codes out of their control, i.e. TLM platform procedures called when their own code access IP registers. Stack frames can be quite confusing as well because it may not be easy to locate the frames at the exact spot where the software really starts. Second, software developers may not see all of their threads if their codes are multi-threaded. The reason is that their threads are embedded in SystemC threads, which may not be visible to debuggers, e.g. the current case for OSCI runtime. Today, debuggers are not much adapted yet for certain non-standard environments such as multi-thread wherein hardware and software simulations are mixed.

Despite all these pitfalls, most of our low-level software runs perfectly well in the native execution environment. In fact, such pitfalls or constraints appear mostly in very high-level software that will be discussed later on. In a nutshell, native compilation is a simple method to start working out low-level codes. It also assists in rising code portability because the same code should run in cross compilation as well where no such constraints apparently exist.

3.4   Examples of TLM-oriented Device Drivers

Without any practical examples, all the approaches described earlier could probably be too theoretical to digest. Let us zoom in on the details of some low-level software already running on TLM platforms through our development work.

3.4.1   SPI Controller Test

The Synchronous Peripheral Interface (SPI) is a very popular protocol widely used in the industrial environments to enable data exchange between a micro-controller and an external peripheral. Instead of plugging a given peripheral directly on a system bus, it is much easier to connect them through a serial interface whose major advantage is the reduction of communication pins. The SPI protocol is founded on the data exchange initiated by a master to a slave at a clock rate determined by the master itself. At each clock signal, the slave must be ready to receive a bit and send out another.

SPI controller tests involve two strictly distinct parts: testing SPI master and/or slave. Data exchange is the principal of testing SPI controller. A fixed set of data must be provided to the SPI controller for exchanging between the master and slave sides, the aim of which is to validate the SPI behavior.

Let us take a closer look at testing an SPI master role (SPI slave role will have a similar testing line). In such test, no SPI slave device is utilized. Instead, it is replaced by a file containing data to be exchanged with the SPI master. The SPI master is exercised by software actions; it also receives the input data from another file holding exactly what it expects to receive. When the software sends a data item such as a byte or something larger to the SPI master, the TLM IP of SPI master controller will read the next data item potentially being sent from the data file representing the SPI slave device. The data read from the latter will then be placed in the registers of SPI master as if it was received in the real situation. Depending on how the software is programmed, the TLM SPI master may update its registers after storing this data.

By comparing the data received from both master and slave sides (more precisely, from their respective data files), the tests of sending/receiving SPI data are carefully conducted by the TLM SPI master controller. Complex data exchanges can certainly be set up, for instance, those including DMA or end-of-transmission interrupt. The validation of SPI data exchanges helps to justify not only the correct functioning of the IP, but also helps to verify the right software programming of the IP registers. With a successful validation of SPI data exchanges, the same software should result in the same test behavior on the real hardware IP that is available later (provided that the SPI slave is correctly simulated by the data file).

Testing a given IP is unfortunately not only limited to its functional tests especially when the IP is synchronized with the external world. In particular, it is impossible to test if the clock programming fits in as required by the slave since the test is not timed. The IP test set will be incomplete if there is no synchronization between the master and slave. If the master acts too fast, the slave will not be able to respond in time. However, the master will still sample the data line coming from the slave to deduce the value transmitted by the slave. This deduction could be incorrect if the timing is wrong. Although the timing programming may be validated statically on an untimed TLM platform, this may not be sufficient.

For that reason, there are two conditions to fully test an SPI IP. First, a timed TLM platform is most of the time compulsory. Second, a mechanism allowing the simulated slave to analyze the timed master responses is required for validating the correct timing of the master. This example illustrates how and when different TLM platform implementations should be employed for various purposes.

3.4.2 I²C Controller Test

The inter-integrated circuit (I²C) bus is a bi-directional two-wire serial bus providing a communication link between integrated circuits. The main difference of I²C from SPI is that I²C supports multi-master mode: I²C allows multiple master devices to connect on the same bus to start the communication at the same time. The collision is resolved electrically, and only one master remains the master of the communication.

I²C is much more complex than SPI. SPI slave is equivalent to SPI master except for the clock generation, whereas I²C slave and I²C master exchange control information such as address, acknowledge, and start/stop right on the bus. This is a sophisticated feature needing software for testing.

For this reason, it is essential to have at least two devices on the bus for testing I²C: a master and a slave. TLM platforms normally provide a single I²C controller that represents a single device on the bus, which supports either multi-master mode or exclusive master-or-slave mode. Unlike SPI controller, the second I²C device cannot be replaced by a file because the control information exchanged on the bus will not be tested. The simplest solution is therefore setting up another I²C controller on TLM platforms *exclusively* for testing. It generates and validates the required bus control information, and its mapping is done on unused addresses.

Once the two devices are properly set up, the I²C controller test can start from any mode. The major challenge of such test is to synchronize the test software precisely between two similar collaborative IPs. While sending information from one of the IPs, another IP must be controlled for its correct receiving of whatever previously sent by the first IP. Polling is not a good tactic because both IPs must be polled concurrently, but interrupts from either IP could arise in any order.

The testing schema describe above is insufficient to test all I²C features. For instance, the feature of *master arbitration lost*[1] in the multi-master mode can only be tested when both IPs agree to set up the same testing condition. Then again, this set up cannot be done by regular register I/O. The same problem may arise in testing all communication errors such as non-acknowledge testing.

Just like SPI controller test, I²C controller test is capable of testing many interesting functional features of the IP before the hardware is ready. It helps to show that the platform runs correctly under standard conditions. This is essentially the first step towards getting a validated TLM platform for running real software programs, particularly real device driver codes.

3.4.3   PrimeXsys UART Linux Driver

The ultimate goal of TLM platforms is *not* developing test software for IPs, but running actual device drivers that will be used on the real hardware platform. Certainly, all events especially errors cannot be triggered to occur as exactly as under real-life conditions. They will however be tested under standard conditions, thus representing the actual behavior most of the time.

An excellent illustration for this concept is the behavior analysis of a real device driver running on a TLM platform. Theoretically, a device driver should run correctly without any modification in the cross-compilation

---

[1] A feature with an I/O starts functioning as a master, but the I/O will become a slave when another master wins the exclusive access to I²C bus (Arbitration).

mode. Let us study the Linux device driver for the UART on the ARM PrimeXsys platform. This driver is initially compiled with the rest of Linux for the ARM PrimeXsys platform. It is then booted on an untimed TLM platform without any modification but some additional error messages for testing purposes.

The experiment shows several interesting side effects. First, the output of messages is very slow. The usage message of *ls* command takes longer than the booting of Linux kernel to display. A closer examination reveals that the DMA is not configured by default for the UART even if the codes are identical. The driver is functionally correct except that an important feature, i.e. DMA, is missing. Although coded, this missing part is not visible enough on the real ARM PrimeXsys platform. In contrast, TLM platforms manage to "amplify" this problem because the missing DMA changes the behavior of TLM platforms significantly. Software using buffered C runtime stdio output routines such as printf, putc, and puts, running on TLM platforms with DMA display a message per transaction while those without DMA () display a character at a time.

Once the problem is fixed, the DMA is enabled in the platform. Yet, the performance still does not show the expected results. All messages are output very quickly but a noticeable delay occurs between usage messages of the Linux *ls* command. Another problem is then identified in the TLM code of UART IP. The added delays in the output for simulating the programmed UART baud rate actually slow down the entire simulation process of UART driver, the reason of which is the ARM platform has nothing else to do but displaying messages. By removing these delays, the UART driver can finally give satisfactory performance results. Indeed, this "discovery" is interesting because time characteristics must be simulated (even on untimed platforms) but with flexible and careful adaptation.

To sum up, running the UART Linux driver on the ARM PrimeXsys platform demonstrates the following benefits of TLM platforms:

1. Device drivers can run without any modification in cross-compilation.
2. Missing performance features can be detected without any special tests on TLM platforms (e.g. DMA).
3. Poorly coded software is immediately revealed by running real software on TLM platforms.

3.4.4 Native Device Driver

Device drivers are not only coded for running in the cross-compilation mode. Running them in the native mode can be equally beneficial for software developers as long as certain coding rules are well respected.

Performance and code portability are the two chief advantages. Software programs can run much faster in the native mode than in the cross-compilation mode, hence higher performance. They can also be compiled on a machine with different constraints such as data alignment, byte order, and language basic types to increase the code portability.

Embedding the software codes of TLM platforms is a very distinctive characteristic of the native environment. It obliges the respect of the host execution rules, including reserved addresses, dynamic loading, name space pollution, etc. These obligations can be very tough barriers to deal with when developing huge software pieces. It is not straightforward to execute both software programs and TLM platforms nicely in the same environment. Beware that all these problems may arise during a project, although huge amounts of code have already been ported in such environment.

The first rule for developing native device drivers is to facilitate the contact point between software codes and TLM platforms. It is usually achieved by using IP register accesses. Such contact in cross-compilation is simply the simulation of a foreign instruction at a given I/O address, which is translated by the ISS into a TLM transaction. TLM platforms just need to issue the I/O and the corresponding results will be given back to the software by the ISS during the simulation of the same instruction. The software in native compilation, conversely, must issue the required TLM transaction by itself. Therefore, the compilation of IP register accesses will need to be transformed into a TLM transaction at the lowest software cost.

Wrapping in macro register accesses is recommended as a good software coding practice. It is particularly useful for increasing software portability onto those systems needing special instructions to access I/O spaces such as I386. Such macro wrapping facilitates the definition of a separate set of macros for the native mode, hence leading to highly portable software. The macro wrapping cannot be applied if register accesses are coded as mapped address dereferences. The reason is that the address range in this case is more likely forbidden to be used in the host execution environment. The initial solution is thus code modification, which may entail additional time delay in the software development exclusively for native compilation.

Another point of attention is accessing the memory shared between the software and hardware for data representation. When everything is ready to be analyzed in an IP, the hardware normally expects to download from the shared memory some data that is already formatted by the software. A good example is a DMA scatter/gather list. Such data representation is a real complex problem because:

1. The simulated hardware may have different byte ordering from the host system.

2. The simulated hardware may align data in a way incompatible with the alignment rules of the host system.
3. The software may use different data types for cross and native compilations, e.g. the "long" type of C language.

There are several good practical rules applicable to solving this situation nicely, which are similar to those used for IP register accesses:

a) Always use fixed length data types so that the field length definitions are not ambiguous, e.g. "int32_t" type of C-99 language.
b) Always access shared data with macros that can be redefined correctly in case of incompatible byte ordering or alignment.

Name conflict is a much tougher problem to solve. Fortunately, it is not something that happens very frequently. This sort of conflict occurs when the software uses an external name that is already defined by the TLM platform. By some chance, the link-editor may detect this as an error. There is however a slight risk that the link-editor may merge it quietly at the same location in the common data segment. That will very likely lead to concurrent use of the same memory location by two modules without relationship. It is then easy to imagine the kind of errors provoked by this bogus situation. Such problems can arise either in static or dynamic link-edit where search results of external names are often hidden by high-level functions..

Sadly, there are not too many solutions for this problem. To cope with it, avoid such naming conflicts by prefixing (or using different name spaces) the external names and minimize using global variables. Name conflict is not a problem directly related to TLM platforms, but it is often encountered by software designers developing huge software pieces.

Concisely, a very important point here is that TLM imposes good software coding practices to prevent some tricky problems from happening during the earliest stage of the development.

4. TLM-ORIENTED OS/FIRMWARE

4.1 Introduction to OS/Firmware

Recall Figure 4-4 shown previously, the software family located above device drivers covers OS (Operating System) and Firmware.

OS is a higher-level software family responsible for integrating all lower-level software pieces to set up a coherent view of the hardware management. Such responsibility is generally entitled to Operating System or Executive

Runtime, which presents a programming interface to higher-level applications.

A key difference exists: Operating System shares CPU time between a large variable set of tasks that are scheduled only when they have something to do, whereas Executive Runtime shares CPU time between a small fixed set of tasks that are called at regular intervals for testing event occurrences and performing potential job. A common point between them is the ability to manage the conflicts of resource accesses for an optimal use of the hardware platform.

Operating System is essentially a piece of complex software for managing task preemption and switching, hardware interrupt dispatches, collaboration between low-level device drivers, etc. The task management role of Operating System is even more distinct when it provides real time functions. Executive Runtime, on the contrary, is considered as a simple task scheduler that neither has potential preemptions between tasks nor interrupt handling; and it has virtually no overhead for task switching. Both of them are of course relatively far from each other in terms of functionality, but they will be considered and described collectively as a single entity called *OS* to cover the two task areas aforesaid.

Another group of interesting software in this family is *Firmware*. It is the software piece responsible for driving some processing parts embedded on the hardware platform. Firmware usually receives and manages specific jobs from an external entity. It therefore plays a mid-level role by unloading some jobs that can be managed locally from the CPU. Such role can be endorsed by running some software on a digital signal processing (DSP) unit to control certain IPs directly for high-level data exchanges with the CPU.

4.2   Purposes of TLM in OS/Firmware Development

Using TLM platforms throughout the development of OS/Firmware serves an important objective: it integrates all lower-level device drivers and executes them in parallel to detect software (potentially hardware) problems related to the interactions of multiple data flows. Such problem detection is either a direct mode by sharing a device driver between two data flows, or an indirect mode where the activities on a data flow prevents another data flow from being correctly managed.

4.2.1   Integration Test Suite Development

As discussed earlier, unit tests are developed for testing a given IP individually (two IPs are required occasionally). Such tests are much limited to testing a single IP without testing the rest of the platform IPs simultaneously. If all unit tests of a platform are pulled together, it is possible to set up an integration test suite provided that the unit tests are developed to run as either standalone tests or concurrent tests in common with other tests.

Therefore, it could be quite easy to set up an integration test suite based on the available unit tests. For a given IP, all of its unit tests can be serialized to form a set of unit tests reserved for this IP. Such unit test set for all IPs can then be run collectively at the same time to exercise all IPs concurrently, hence leading to a proper integration test suite.

This approach, nevertheless, is not that straightforward for untimed TLM platforms. Due to their untimed characteristic, setting an I/O transfer via a register write is virtually immediate. Thus, no I/O parallelism can have effect in the set of IPs under test. Moreover, an interrupt will be triggered as soon as a register access is completed if the interrupt mode for I/O completion signaling is used. It is possible to chain all these tests in order to run them one after another. Yet, this is still not quite a real integration test suite.

The missing part is an executive runtime that can run one test after another. It should avoid running the entire test set of a given IP right after running the entire test set of another IP. A better way to handle this is running interleaved tests in order to exercise all IPs more frequently.

On the contrary, an integration test suite can be set up very nicely on timed TLM platforms. The reason is that each I/O consumes some time to signal its completion, and that consequently allows running multiple tests for different IPs in pseudo-parallelism. The term "pseudo" signifies the fact that I/O completion time is accurate but its progression is without cycle accuracy.

Integration tests should also be in charge of testing arbitration, i.e. hardware conflict resolution. The typical examples of such conflict are two concurrent interrupts or I/O bus accesses. Interrupt conflicts can be validated on timed TLM platforms whereas bus access conflicts can only be tested on BCA platforms.

In brief, integration tests are merely some test set on untimed TLM platforms; they provide much more interesting results on timed TLM platforms; and finally give solid outcome on BCA platforms. Therefore, TLM platforms should be considered as initial test platforms where integrations tests are executed and debugged before other further accurate platforms are made available.

4.2.2   Quick Application Software Evaluation

An attraction for using TLM platforms in OS/Firmware development is the *quick* set up of a complete system integrating hardware and OS, which aims at evaluating external higher-level application software. Throughout a new SoC development, it is always a challenging mission to validate if the design meets the software requirements until the software really runs on it. The early availability of TLM platforms allows software developers to get a precise image of the final hardware platform for running software ahead of time.

Today, the standardization of interfaces and low-level services such as Windows, POSIX, OSEK, and iTRON has facilitated the implementation of such interfaces on top of a number of Operating Systems. As a result, higher-level software can be ported much easier from one platform to another. Combined with TLM platforms, these interfaces and low-level services offer high-level software developers a complete system that is ready to support high-level software. Depending on their levels of accuracy, TLM platforms serve extensively as evaluation systems that support major OS available today.

A complete system consisting of OS, device drivers, and TLM hardware platforms is essentially the very first integrated system accessible to high-level software developers. Such a complete system holds several important characteristics as follows:

1. It is not restricted for large deployment since it is purely software; the number of systems available for using is thus not limited to just a few fragile hardware boards.
2. It implements a realistic system platform whose accuracy depends on the accuracy of TLM components and the software integrated.
3. It provides a platform with a coherent behavior of all integrated platform parts, i.e. both hardware and software.

4.2.3   Closed Integrated Software Module

During the development of OS/Firmware, TLM platforms also serve the purpose of employing black box tests and pure binary software codes.

TLM platforms are established as accurate representations of some existing or upcoming real platforms Thus, they must support binary software codes intended for running on the real platforms in a transparent and reliable manner. Software developers count a lot on this feature to prove not only the accuracy of TLM platforms, but also the correctness of their software with respect to the real platforms.

Note that this particular characteristic is only necessary for cross-compiled platforms because it is worthless to set up binary software for some platforms that will never exist. A prototype can hardly provide the same feature described here because it will never be accurate enough to run binary software codes as a black box test. This is also hindered by other reasons such as netlists, definitions of IP registers, component timings, etc.

To conclude, running pure binary software codes on TLM platforms have two ultimate goals:

1. Validate the TLM platforms when the real hardware already exists with the same software.
2. Validate the software provided as binary codes with some extensions, which are developed for supporting additional hardware features on a new platform compatible with some existing ones.

4.3   Approach to OS/Firmware Development

OS or Firmware developed for running on TLM platforms are not directly related to the hardware simulation. They should consequently be less sensitive to certain low-level details implemented in TLM platforms. Bear in mind that OS/Firmware is a special software layer responsible for the employment policy of the mechanisms defined by lower software layers. This is exactly where meticulous care must be taken to handle the capabilities of TLM platforms correctly. The approaches to developing OS/Firmware mainly focus on how to get a complete hardware/software system to run efficiently without wasting simulation performance in useless tasks.

4.3.1   Active Waiting Loop Avoidance

The foremost software quality is being able to utilize the underlying hardware at the optimal level. This is nevertheless quite a tricky game to play with on a simulation platform such as TLM. Since the hardware parts of a simulation platform are merely some simulated models with various accuracies, certain programming techniques may appear inefficient in the simulation run although they may work reasonably on the real hardware. The main reason is that some trade-off between performance and accuracy must be made on simulation platforms.

The active waiting loop is an example of the programming techniques difficult to be adapted on simulation platforms. On the real platform, such software will loop perpetually through a list of awaited events until one of the events finally occurs. The same software technique runs in the same way on TLM platforms but less efficiently, as the hardware event will only occur when the software allows it to occur. The software run indeed prevents the parallel run of TLM platforms, which is supposed to raise the subsequent event waited by the software.

This situation is similar to a deadlock but not fatal because the software might be preempted sometimes. Rather, active waiting loops are considered unproductive since the system cannot evolve during their execution. Keep in mind that events are driven by the hardware speed on the real hardware platform for which the real time is continuous, whereas events are driven by *no* previous activity on TLM platforms for which the simulated time is discrete.

Therefore, it will be wise to have useless software suspended until the next hardware event occurrence. The software should then loop again until it finds the occurred event. Since such software loop is sensitive to hardware events, it is not easy to program it transparently for either the real or TLM platforms.

Actually, new requirements for low-power consumption on SoC have helped to solve this tricky problem. Under this concept, any software with nothing constructive to perform will simply switch to the low-power mode to wait for the next event. Such switches are handled by some hardware interactions on TLM platforms, which can subsequently advance the system to the following event in line. Essentially, this is what will really occur on the real and TLM platforms. The low-power feature therefore avoids the active waiting loops and enables the same binary software to be used on both platforms with equal efficiency. Without the problem of active waiting loops, TLM platforms are once again ready to drive software towards the better use of the underlying hardware.

4.3.2   Hardware Interrupt Management

Interrupt management is another problem similar to the active waiting loop. Normally, starting an interrupt I/O on the real platform takes some time. The software will not just wait for the completion of the I/O event by doing nothing. Instead, it will try to perform some other useful jobs while the hardware processes the I/O work. This is feasible for the software only if the hardware takes long enough to notify interrupt events. As discussed earlier in the interrupt management for device drivers, receiving interrupts too early could be unfavorable as the software may not be able to perform other useful jobs or may even spend more time in handling the interrupts. This is particularly true for the management of input device events arriving at unknown (or very high) frequency.

Looking at the whole picture of a system design, a given software should run equally well either on the real or TLM platforms. Similarly, polling and interrupt modes implemented in the software should be valid for both platforms despite their different performance behavior.

Recall that OS/Firmware involves more than a single device driver. There is a potential side effect of using immediate interrupt notices. It may happen that the software reads data from a hardware source that always acquires ready to-be-consumed data. On the real and timed TLM platforms, acquiring data consumes some time that the software can run other tasks while reading data. However, on untimed TLM platforms, it occurs that interruptions related to I/O completion are somewhat instantaneous following their respective I/O activation, which may lead to a kind of apparent execution starvation for other processes. Polling is therefore the better solution for handling interrupts on untimed platforms.

To prevent such misbehavior that the software cannot avoid by itself, certain safeguards must be provided in TLM platforms. These safeguards will serve as the guidelines to standardize the software codes running on different platforms, because software designers develop their software based on the TLM platforms provided to them.

4.3.3   Native vs Cross Execution Environments

Native versus cross software compilations are discussed in sections 3.3.5 and 3.4.4 to describe the compilation nature of low-level software like device drivers. Regarding higher-level software such as OS/Firmware, software developers must consider carefully some different behaviors introduced by its execution environment.

Performance is always much faster on a native platform than on a cross platform. This is the reason why most of the SoC developers tend to use native compilations for their TLM platforms. Such high performance, however, is not as easy to reach on OS/Firmware layers due to certain processor-specific features that are tough to cope with; among them the most noticeable ones are the virtual memory management, the code and/or data cache management, the execution and interrupt paths, which are clearly under the responsibility of OS/Firmware.

Another important concern for the execution environment is the strategy of software debugging. On the real hardware, software developers debug by either plugging in additional hardware pieces to control program execution or embedding software debuggers in the OS/Firmware codes. The most common solution is based on the Joint Test Action Group (JTAG) and some hardware extensions such as In-Circuit Emulator (ICE) or User Debugging Interface (UDI) for a complete execution control, while the latter is based on the low-level CPU control.

JTAG and other hardware-based solutions are not practical for TLM platforms because they are too dependent on the hardware availability. On one hand, embedded software debuggers can only be supported in cross-compiled environments, but on the other hand, they can benefit from the precise debugging information via a debugging server that can stop ISS at the granularity of an instruction. This is relatively much easier compared to the real platform on which software debuggers must trap current execution flow on the real processor by exception, hence causing an intrusive debugging process. The direct impact of these differences is that the debugging strategies might not be similar between the real and TLM platforms. For native platforms, the only debugging solution is using host debuggers with some proper adaptations as explained earlier in section 3.4.4.

4.3.4   Virtual Memory Management

An important software aspect dedicated to OS/Firmware management is the Virtual Memory Management, namely the handling of the mapping between physical pages (the ones actually in memory), and their association in the standard addressing schema. This is usually achieved either via a Memory Management Unit (MMU) or via a TLB Translation Look-aside Buffer (TLB), internal to the CPU.

Under the cross-execution, the ISS is responsible for such management, and the TLM platform only receives accesses to physical addresses. This is however a costly management as each memory access must be translated from virtual to physical addressing space, which relies on search tables. The main advantage is that the accuracy of the management is high, and that all MMU-related traps are notified to the OS.

Since performance is the main problem of memory management, then it could be useful to take native compilation into consideration. The main problem now is to get the maximum possible accuracy in order to be in a simulation environment providing enough realism for software development.

The first thing to remember is that the embedded OS and firmware usually have nothing to do with on-demand page swapping (i.e. the ability to put unused memory pages on secondary memory such as disk drives). The memory management is restricted to the ability of running multiple applications within a finite amount of memory, which is more of a memory placement problem rather than virtual memory management problem: if all of the applications are not able to fit into the available memory, then the system cannot work safely.

This is the reason why for OS/Firmware coming with source code, the most efficient approach seems to replace virtual memory management by the dynamic placement of all the processes in the virtual memory of the TLM platform execution. The main advantage is the performance gain, as there is no overload of translating virtual addresses to physical ones (the code and data addresses are relative to a register and allocated memory areas being directly accessed). The main drawbacks, on the other hand, are the non-protection of a process data against erroneous accesses (by itself or by other processes).

This solution can be applied with Linux, for which µClinux is the right and most efficient solution. The reason is that OS provides the same API not only to user-mode applications, but also to dynamically loadable kernel modules such as device drivers, file systems, etc. Such pieces of software may be almost completely and transparently tested and debugged within a MMU-less environment, and then integrated into a cross-integrated TLM platform with complete MMU support for final validation.

4.4   Examples of TLM-Oriented OS/Firmware

The current section provides a brief description on some OS/Firmware examples already running on TLM platforms through our development work. These practical examples illustrate how some approaches described earlier are used.

4.4.1   Quick Setup of Integration Test Suite

As soon as a TLM platform is set up, the immediate subsequent step is running unit tests for every IP of a platform. These unit tests must be conducted IP per IP to evaluate the correct functioning of each IP. Once the first step succeeds, the next step is running integration tests to validate the integration of all IPs in the TLM platform. This step is particularly important for validating the areas that necessitate arbitration for handling concurrent accesses to shared resources such as bus or interrupt controller.

Since parallelization is required, an integration test must be written to deal with more than a single IP at a time. Moreover, integration tests need a test harness to optimize the pipelined execution of all IP unit tests and to exercise all IPs in parallel on a given platform. The test harness referred to here is indeed an Executive Runtime that schedules each task to manage a single IP. It may also include interrupt management to validate the right functioning of the shared interrupts. Another advantage of the Executive Runtime is its ability to handle multiple IP tests such as the I^2C controller test described in section 3.4.2. Test codes for each platform IP is placed in a different task scheduled by the Executive Runtime in order to retain the paradigm of managing an IP per task.

A test set should be arranged to complete properly because an Executive Runtime does not really stop by itself. It is vital to ensure that there is definitely nothing else to execute in order to terminate the integration test correctly. A good practice here is to let the underlying SystemC runtime "discover" the simulation completion by having no more events to handle. In this way, the simulation will exit without returning to the software; hence avoid having the Executive Runtime to deal with the test termination.

4.4.2   OS on ARM PrimeXsys Wireless Platform

The ARM PrimeXsys Wireless Platform (PWP) is an extendable development platform whose description is open to the public[2]. Software developers can understand this complex platform much better by running a simulation on the TLM platform of ARM PWP.

Booting an OS on PWP aims at verifying the software behavior on both simulation and real platforms. It could be really challenging to boot an OS binary image on a cross-compiled platform if a bug appears in the OS code execution. The core architecture assembler is the only accessible debugging level, which is unfortunately not so obvious to get information from. Bugs found in this type of simulations are usually related to some subtle behavior differences between the real and simulated hardware. If the source codes are unavailable, tracking such bugs becomes much tougher.

The first right approach is to run an OS whose source codes are available, for instance, running Linux on PWP. In general, an ISS-embedded debugger is aware of the OS object format and thus capable of conducting symbolic debugging. If the debugger recognizes the structure of the internal threads, the whole OS execution structure can be exposed through the debugger interface. As a result, the debugging process becomes much easier to control. This process is a little more difficult in the case where MMU is handled by Linux because the debugger must understand the virtual memory translation. In this case, it is easier to start running the platform with uCLinux, i.e. the Linux without MMU port. Once the PWP TLM platform is extensively exercised by this OS, it is time to get some binary OS such as Windows or Symbian running on the platform.

Booting OS on TLM platforms is advantageous. It gets ready a complete platform for software developers to design high-level software. In addition, it helps to find missing or incorrect OS codes that are virtually invisible on

---

[2]   Available at http://www.arm.com/products/solutions/PrimeXsysPlatforms.html

native platforms. A real-life example is the public code of PrimeXsys Linux for the ARM926EJ-S platform. By comparing the TLM simulation to the one of real platform, two errors are found in the public code. We notice that the bootstraps for cache enabling as well as the UART accesses to DMA are too slow on TLM platforms while they appear apparently correct on the real platform. This example shows the direct advantage of TLM simulations.

4.4.3   Native OS Emulation on Video Platform

The video platform is an interesting platform type to learn more about TLM concepts. As the video flood flows from one IP to another, the central processor only acts as a director without a straight view of the video flood. IPs are generally complex hardware pieces that run some firmware on top of an internal micro-controller.

Simulating precisely all IPs on a TLM video platform is not quite feasible because not all IP models are made available quick enough for software developers. It is possible to build accurate TLM models of these complex IPs and integrate them as a single TLM platform. The resulted performance, however, is very likely to be slow. Bear in mind that one of the primary goals of video platforms is the pipelining of the video flood. Video components are therefore designed to run in parallel. Each of them must also run its associated code accurately, and that will consume many CPU resources. Consequently, the sum of such consumption will result in a very slow platform.

Running software on a simulated platform is nevertheless still very important for software debugging. A different approach must be adopted to cope with the problem of slow simulations. The proposed idea is splitting a given platform into simpler but meaningful hardware pieces for software development.

In other words, if the software perceives a specific IP block and its associated firmware as a black box, then a single TLM model is conceived to represent the whole IP block including the associated firmware. Sometimes, software developers may need to debug both software and firmware on the same platform. In that case, software developers must set up the platform in such a way that the internal IPs and their associated firmware are exposed.

The benefit of such platform setup is the close simulation of the IPs whose associated software requires debugging, while maintaining the overall platform performance at an acceptable level through keeping other IPs as efficient black boxes.

Native emulation is the suitable solution for working with the firmware of complex IPs because it retains both performance and functional accuracy of IPs. In addition, debugging the whole platform in native emulation allows grabbing an image of the OS running on the central processor and the firmware running on the IPs of interest. Therefore, it is feasible to analyze the interactions between all the software running on the different IPs at the same time, or even in the same debugger session.

Having TLM models as black boxes of IP hardware and their associated firmware may have some impact on the OS drivers managing these IPs. If the firmware is encoding/decoding a well-established algorithm such as MP3 or MPEG4, it will be quite straightforward to set up an IP model running the similar code but with different interfaces. There are two solutions for the OS driver:

1. adapting the driver to the simulation interface, i.e. a fast solution;
2. adapting the simulation software to the IP interface, i.e. an accurate solution.

Both solutions are valuable approaches that should be applied at different phases of the SoC design process, depending on the expectation of the software development. The accurate solution, however, should be retained once it is available because it allows debugging the actual OS driver.

5.   TLM-ORIENTED APPLICATION SOFTWARE

5.1   Introduction to Application Software

Although the two lower-level software families i.e. device drivers and OS/Firmware bring interesting results, the ultimate goal of software developers is to get the final application running on TLM platforms as soon as possible. This high level is probably the easiest to set up on TLM platforms because it has very little or no relationship with the actual hardware programming.

Building and running a complete application on TLM platforms is an ambitious objective to achieve. In common practices, applications normally run on a prototype of hardware board whose central SoC is only a part of the board. Today, the implementation cost of hardware prototypes is skyrocketing due to the explosive SoC complexity. Running applications on a much more complete platform like TLM is therefore getting increasingly attractive for SoC developers. The TLM platform is not only a hardware platform that is accurate enough and available significantly earlier, but it is also able to easily outperform the equivalent RTL platform.

By running the critical software parts interacting frequently with the lower-level software, software developers should gain enough confidence that the application is ready to be integrated as soon as the prototype of the final platform is available. The critical software parts already layered and packaged as independent libraries such as protocol stacks, data stream decoders, all sorts of data filters, graphical environments are among the most essential parts to be verified in this manner.

5.2   Purposes of TLM in Application Development

5.2.1   Final Validation Test Development

Once integration tests are completed, validation tests can be started. A validation test exercises a given platform in the environment that it is specifically designed for running in. The principal idea is to define a set of representative test scenarios intended for the final platform execution, i.e. the highest level of tests that a platform must go through for its validation.

Validation tests are crucial in assuring the accuracy of the integration of TLM platforms. These tests aim at demonstrating that TLM platforms behave exactly as they are designed. If TLM platforms show their high fidelity to the hardware platforms, the same validation tests should provide the same level of confidence on the real hardware platforms.

Since accuracy is a characteristic that becomes more critical close to the end of SoC development, a validation test will emphasize more on the accuracy of a platform than on its performance. For this reason, running validation tests can sometimes be very time consuming. The focus of validation tests is on the whole software rather than on the TLM platform because the objective is to try out the TLM platform in the real environment.

It is vital to have validation tests running in a simulation environment that is as accurate as possible in order to test the internal behavior of the platform. Thus, TLM platforms must interact with the external world at the highest possible accuracy. This requirement is less strict for the previous two lower-level software families because their main purpose, i.e. testing TLM platforms, is different from the one of validation tests for high-level software. As such, validation tests can be considered as part of the interoperability tests.

5.2.2   Performance Experiment

While the OS/Firmware is sufficient to demonstrate that a platform is functional by itself, the application software is intended for providing additional feedback from TLM platforms. More precisely, it places TLM platforms in a real application environment wherein non-functional results can be obtained from the whole system.

Such non-functional results can be considered altogether as performance results, covering timed profiling (e.g. latency, speed, throughput, deadline), untimed profiling (e.g. counter, contention, bottleneck), and other factors such as power consumption estimate, security evaluation, fault tolerance, resource footprint, etc. Without TLM platforms, it is almost impossible to assess these performance factors accurately since the whole application will not run in a realistic environment.

Running application software on TLM platforms is a good occasion to conduct a brief benchmarking for the internal behavior of a platform or an IP, i.e. behavior due to the fine-grain impact from certain hardware features on the platform. Transactional analyses in this realistic benchmarking serve as accurate sources for significant decision-making in SoC development. To perform such accurate simulations, timed TLM platforms are compulsory.

5.2.3   Impressive Demonstration

Running high-level software is not only useful for hardware and software developers, but it is also valuable for other professionals involved in the SoC project. A lively demonstration of a given application can be surprisingly rewarding for marketing crew as well as final users.

An impressive demonstration of the whole system is the real foundation of communication for marketing crew. The demonstration is intended not only to prove the platform compliance with the requirements, but also to illustrate the impact of pulling all the requirements into the entire system. Final users, on the other hand, can start validating if the platform fits their design requirements without waiting for the first prototype. If the result is negative, there is still ample of time margin to modify the platform before the real hardware advances. After all, TLM platforms are simply some software pieces that can be easily altered.

5.3   Approach to Application Software Development

This section provides some general advices to develop application software revolving around TLM methodology.

5.3.1   Provision of Realistic Environments

Developing a specific application targeted at an embedded system on a real chip is quite a different matter from developing the same application for running on a workstation or application server. The reason is that the embedded system is a *real* environment with plenty of constraints in terms of computing resources, bounded memory size and less powerful CPU. On the other hand, the workstation or application server protects the software developers from this realistic environment with their own software environment. Such *unreal* environment may tolerate certain programming pitfalls that could become threatening on the real embedded system. Therefore, application developers must be aware of this potential discrepancy and design their software tailored for the optimal use of the computing resources.

A significant part of the application software can be developed natively on workstations or servers without TLM platforms. Although well fitted on this untargeted foreign environment, the resulted software often triggers a disaster on the target platform. Several reasons can explain this situation. In general, the larger the SoC project, the more difficult the anticipations will be. Collected below are among the most common examples:

1. non-executable application due to the excessive memory consumption by the application itself on the target system;
2. system crash due to some leakage in the resource consumption;
3. inefficient codes due to a slower processor;
4. deadlocks of multi-threaded applications due to different scheduling.

Looking at all these potential problems, it is absolutely critical to get ready a realistic environment for embedded applications as soon as possible. Through a realistic environment, such problems can be detected in the early phases of the application software design ranging from development to execution and simulation. TLM platforms offer a 3-in-1 solution that covers the three environments for bringing realistic effects, i.e. development, execution, and simulation environments.

The *development* environment is the set of applications and libraries necessary for building and debugging a given embedded software. Its early use within the application design process enables the early detections of compilation problems, missing target library functionalities, and resulted image size to be loaded on the target platform.

The *execution* environment is the set of resources involved in the embedded software execution. This is the heart of what TLM platforms have to offer to software developers. Running applications on such platforms give software developers a realistic idea of any potential execution problems hidden in their applications.

The *simulation* environment is the set of external devices connected to TLM platforms. Its mission is to provide software developers with a realistic external simulation that will exercise the whole system hardware and software. Essentially, this environment simulates the application in the setting that it expects to run in. Such simulation is important for observing the realistic aspects of the embedded system that depend very much on the data exchanges with the external world.

5.3.2   Attention to Application Performance

The amount of software to be run at the application level is huge. It is very likely to run on many loosely coupled IPs and split into a central application software with multiple firmwares around. Therefore, performance becomes a core issue for using TLM platforms in order to extract quickly interesting results. If the accurate timing is one of the major expected results, native compilation is of little help.

Running multiple loosely coupled software pieces in parallel provides software developers with the ability to parallelize them easily. Although platform IPs may seem independent from the angle of software, TLM platforms impose a certain level of serialization among them because they interact altogether at the hardware level. Consequently, a very accurate TLM platform from the hardware perspective may become much less efficient in running an application software. It is therefore extremely important to choose the accuracy level of each TLM component very carefully for a system integration.

Nevertheless, keep in mind that the host processor is usually much more powerful than the simulated hardware (for the majority of the simulated IPs except the central processor). The global performance of the TLM platform thus correlates with the ISS performance. Such correlation is also true for IPs running a firmware on a little embedded micro-controller turning at low speed. If the TLM platform embeds an ISS for this micro-controller, then the ISS performance will be high thanks to a simple code emulation as well as the high relative speed between the micro-controller and the host processor. Watch out: the firmware running on this IP could be an active loop waiting for an event to be raised by another IP, and that may cause the system temporarily doing nothing constructive for the platform evolution. Of course, this is another matter on a timed TLM platform as the relative performance of IPs are already taken into consideration.

To conclude, the challenge of running an entire application on a platform may turn out to be much tougher than expected. The challenge has no close relationship with the hardware, but rather in validating the integration of all software pieces. It may lead to coherent yet antagonistic directions from the point of view of application performance.

5.3.3   Control on TLM-Specific Code Amount

TLM and real platforms provide software developers with different but complementary advantages. TLM platforms, however, are not set up to run huge applications due to a costly price to pay for the high accuracy, i.e. low performance. Yet, there is still a real interest to run huge software slowly on

TLM platforms: obtain a level of internal observability that is almost impossible to get from a real platform.

When a TLM platform is stopped, all of the platform components are stopped at the same time in the same state even though they are not tightly coupled. This is a very advantageous situation to examine the component states in details. Such convenience is not easily reachable on a real platform because everything is fit into a chip, which is only accessible through some internal complex and indirect tracing mechanisms such as JTAG.

Software developers must refrain from writing too many software specific to TLM platforms. This can be helpful at the early stages of a project in order to make use of native compilation. However, it becomes less and less useful as the project advances because the platform will normally be more and more complete while adding new functionalities. Having such code at late phases might be mainly for catching some subtle bugs. It will not be reused on the final hardware and thus can be considered worthless to be developed.

Bear in mind that TLM platforms are pure software. If the real hardware is available, it will certainly be more efficient to debug an entire application on it than on the TLM platform (provided that lower software layers are sufficiently debugged beforehand on the TLM platform). Although subtle bugs can be discovered faster on the real hardware, TLM platforms still merit a vital role at this stage to continue testing certain software parts separately from the whole system view.

In brief, software developers must *wisely* determine the software amount developed particularly for running high-level software on TLM platforms, knowing that it is preferable to target final applications on the real platforms for the reasons of debugging efficiency and code reusability.

5.4   Examples of TLM-Oriented Application Software

Typical applications running on TLM platforms are those interoperating different IPs through an embedded processor or micro-controller. Particular applications running on a single IP can be interesting for demonstration but not really valuable for debugging.

5.4.1   Multi-Processor Platform Application

A multi-processor SoC platform (MPSoC) is a platform that embeds more than a single processor of the *same* type on which the system workload is distributed. Other main parts on MPSoC platforms include communication channels and potential managers to handle the multiple processors.

MPSoC are extremely complex platforms with more than one CPU to run the software that manages the whole platform, which consequently leads to a distributed management model. Such hardware implementations could be too complex to understand for software developers. Rather, they should try to comprehend the TLM model of the platform and figure out how to get it run. That will focus software developers on the problems related in running an application on these platforms, instead of understanding how the platforms work.

Communication is another interesting point to employ TLM models of MPSoC platforms. The reason is that the complex software running on MPSoC platforms always tries to split its workload on all platform resources for an optimal utilization. It is therefore fundamental to have an excellent control over the communication and data exchanges in order to master such platforms. The unified view and global fine-grain control of each processor through TLM platforms allow software developers to retrieve and analyze the MPSoC platform behavior easily. These analyses can be accomplished without overlooking the subtle platform management normally handled by software such as cache coherence with DMA and shared memory.

Typical MPSoC platform applications are generic parallel applications in which various tasks such as graphical encoding/decoding, network routing, scientific computations can be executed *indifferently* on any processor. The goal of using TLM platforms for such applications is not really running the computations, but rather validating through small examples that the workload is well distributed among all of the processors. It is easier to conduct this sort of validation on the slow but accurate TLM platforms than on the real hardware whose activities are much more difficult to control.

5.4.2   Centralized Multi-Architecture Platform Application

A multi-architecture platform denotes a platform that embeds multiple processors of *different* types. The overall workload is distributed among all of the processors but these processors, unlike MPSoC platforms, do not play the same role. Software must be split into pieces dedicated to each type of processors.

Multi-architecture platforms are usually called for applications dealing with the parallel handling of multiple data flows whose management is centralized on one or multiple processors of the *same* type. Examples for these platforms include telephone, set-top box, and complex audio platforms.

Since each processor is of different type on multi-architecture platforms, heterogeneity and synchronization are unsurprisingly the most difficult parts to master. Indeed, communications in multi-processor platforms can be considered as a particular case of communications in multi-architecture platforms. High-level management applications are responsible for optimal distribution of the instantaneous workload to the right processor at the right timing. TLM platforms provide a real advantage by giving a high-level view of different software pieces running on different processors.

The DSP plays a particular role as a multi-architecture platform. DSPs are not general-purpose processors but they can be continuously reloaded with a new program to perform a new function required at a given time. Debugging DSPs is exceptionally complex because the software running on some of the processors changes constantly. This characteristic could be a potential source of bugs that is not easy to detect and fix. To handle the communication between a CPU and a DSP, software developers need to obtain a coherent view of the software distributed on heterogeneous processors all the time.

5.4.3 Pipelined Multi-Architecture Platform Application

The pipelined multi-architecture platform is another kind of multi-architecture platforms. This is a platform whose IPs are integrated in such a way that a data flow will stream from a specific IP to another specific IP through the whole platform.

The role of the CPU on a pipelined multi-architecture platform is similar to the one of the centralized one, i.e. it organizes the data flow and manages the external events. Typical platform examples are multi-media decoders.

Getting ready the application software as soon as possible for such platforms enables the flow design validation and the early revealing of any potential bottlenecks that may threaten the overall platform performance. These platforms usually consist of multiple IPs with micro-controllers that run a firmware not modifiable until the next platform reset. The overall software is split into multiple pieces that are loaded into their own target IP processors, which are independent from each other except for their synchronization.

TLM platforms can simulate this type of complex platforms for validating not only the IP-dependent software individually, but also the overall application that subsequently allows careful checking of certain platform aspects during debugging process.

6.   CONCLUSION

Running software on TLM platforms leads to splitting software in three layered categories that correspond to the three principal software-testing layers: unit test, integration test, and validation test. For each layer, TLM platforms offer software developers the simulated platform that they need for executing and debugging their software. The choice of the accuracy level of TLM platforms is crucial, as the software will appear to run less efficiently on a more accurate platform. Figure 4-5 recalls the idea of relating different software families and environments in the V-diagram of software testing.

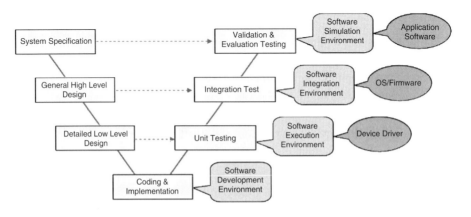

Figure 4-5. Software Families and Environments vs Software Testing

Developing a complete system based on TLM platforms can significantly improve the methodology and the schedule of the hardware and software design. Software developers are able to test their codes on simulated but accurate hardware platform long before the real hardware prototype is ready. Even after the real hardware is made available, TLM platforms continue to bring software developers important information that is not obtainable from the real hardware. Essentially, TLM platforms play the central role in hardware/software development as a common exchange platform between these two development teams.

TLM platforms are also considered as the software bug amplifiers. They reveal a more general behavior of the hardware, which is normally not easily accessible to software developers. This advantage reinforces good software practices for software families ranging from low-level codes to software architecture layering.

In conclusion, TLM platforms shorten the global time-to-market and raise the overall quality of SoC projects. Uncovering software and hardware bugs long before the real platform availability considerably trims down the cost of bug fixing. Last-minute patches can then be avoided most of the time, giving software developers more time to work on performance issues.

Chapter 5

FUNCTIONAL VERIFICATION
From The TLM Perspective

Thibaut Bultiaux[1], Stephane Guenot[2], Serge Hustin[1], Alexandre Blampey[2], Joseph Bulone[2], Matthieu Moy[2]

STMicroelectronics Belgium[1]; STMicroelectronics France[2]

Abstract: Functional verification has traditionally focused on providing tools to generate tests and measuring their so-called coverage. The need to provide the correct reference data has had however relatively little attention. This chapter describes how to apply TLM models as executable functional specifications to generate the compulsory reference data required by functional verification environments. We further explain how these models can be used in conjunction with other verification techniques such as hardware emulators, and how formal verification techniques can be applied to TLM models.

Key words: verification; IP test bench; system test bench; test scenario; input data; test stimuli; expected value; golden reference; data manager; signal-transaction conversion; pin convertor; signal convertor; bus functional model; monitor; checker; transactional co-emulation; formal verification; LusSy; Lustre.

1.   INTRODUCTION

The functional verification of a SoC design is a phase that guarantees the compliance of the design implementation with its specification. It is a complex and time-consuming design step, accomplished by converting the specifications into a combination of:

1. *Stimuli and expected results scenarios*, verifying that the design produces the expected results when applied with the stimuli.
2. *Golden model and stimuli constraints*, verifying that, whatever the constraint-compliant stimuli, the golden model and RTL behavior are equivalent.

F. Ghenassia (ed.), Transaction Level Modeling with SystemC, 153-206.

3. *Properties and stimuli constraints*, verifying that, whatever the constraint-compliant stimuli, the properties hold true.

Using TLM for performing these tasks allows sharing TLM verification scenarios and models between the design groups, hence increasing productivity, integrity, and task parallelism.

Productivity is increased by avoiding the duplication of verification scenarios and models. Through higher simulation speed by using TLM models alone or by TLM co-emulation, productivity can be further boosted.

Integrity is improved, for instance, by assuring that the model used by the software group is being verified against the RTL model employed by the hardware and verification engineers.

Task parallelism is provided by the possibility to develop scenarios and tests using TLM reference model while the corresponding RTL implementation is being developed.

This chapter discusses extensively on how to apply TLM to SoC verification strategies. To begin with, it explains how TLM influences the verification flow. It then describes the important aspects of the verification flow in using the TLM methodology. The strategy for a complete TLM verification flow is also detailed, followed by the illustration of a TLM use model within the verification flow. Lastly, the chapter explains how properties can be proved directly at the TLM level.

2. A NOVEL APPROACH TO SOC VERIFICATION

TLM is put forward as a novel approach to the SoC design residing at the transaction level, which is revolutionary compared to the RTL design at the signal level. The focus of this chapter is introducing TLM as a different approach to the SoC verification flow.

The classical flow of a SoC design stretches from the RTL design down to the post-layout GDS format for tapeout. Throughout the flow, a number of stages are performed. Each of these stages allows a refinement from a certain level of abstraction into another level of higher precision. Based on the RTL design, a logic synthesis is carried out to obtain a gate netlist. As denoted by their names, RTL designs deal with circuits at the register transfer level while gate netlists handle circuits at the gate level. The automated process of logic synthesis can be verified by the method of formal proof.

The TLM-oriented SoC verification extends the SoC design and verification flow to a higher level of abstraction than the classical flow. This new methodology, however, has yet to incorporate an automated process to pass TLM designs to RTL designs. Some studies have been going on in this direction but to no satisfactory solutions so far. To date, the tools developed from the research are not as stable and efficient as those for the conventional logic synthesis. Such TLM-to-RTL refinement remains a manual task in general. The two models are sometimes prepared by two different engineers, making the refinement task a more complicated process.

As natural as using formal proof to verify the conversion of RTL designs into gate netlists by the logic synthesis, it is undoubtedly essential to verify the conversion of TLM into RTL as well. The verification of both TLM and RTL models is crucial to guarantee their equivalent behavior. The software team may otherwise run the risk of simulating a different behavior than what the hardware team plans to implement.

In certain cases, the direction of the TLM-to-RTL conversion is reversed, i.e. the RTL design is developed and tested before the TLM design. In that case, RTL and TLM swap their roles. The RTL model serves as the reference model whereas the TLM model becomes the design-under-test (DUT). Yet, the equivalence of RTL and TLM behavior remains to be proven.

No tools of formal proof are developed for comparing the RTL and TLM behavior thus far. Instead, a special strategy is adopted as an alternative. This is a tool-independent strategy that costs little workforce and can theoretically be automated.

The TLM verification methodology is not only capable of performing something necessary and critical in the SoC design cycle, but also of improving the SoC design flow. Most of the verification strategies prepare a golden reference model, which is very often written as a C model. Actually, only a little extra effort is required to write up a TLM model instead of some C functions. This little effort, however, brings large gain in terms of design reusability. Indeed, TLM reinforces a reuse-compliant verification methodology by reusing a high-level test bench to verify an IP model conceived at a lower level of abstraction. On top of it, the RTL verification becomes easier through the TLM approach, giving designers more room to focus their efforts on the generation of test stimuli. The routine of a TLM-based functional verification is briefly described hereafter.

The TLM model of a given design-under-test, i.e. DUT, has first to be written; followed by the corresponding TLM test bench, i.e. the high-level test bench. Subsequently, the reference model for the design is developed under the TLM test bench. The TLM DUT is simulated by the TLM test bench for a given test scenario and the associated input stimuli. The results of the simulation execution are collected as a set of expected values, which serves as the reference model. Once the golden reference is obtained, the same TLM test bench is reused for verifying the RTL model of the DUT. The RTL DUT is simulated by the same TLM test bench for the same test scenario and the associated stimuli. The resulted values are then compared to the values of the golden reference model in order to perform a functional verification of the DUT. This methodology should be applied along with other integration test benches at the top-level in order to test the behavior of a given DUT wihtin a system environment, for instance, SoC environment.

3.   APPLYING TLM IN SOC VERIFICATION FLOW

In the classical SoC design flow, design engineers need to interpret a written specification document in order to produce a synthesizable code in hardware description languages (HDL). Engineers are often held accountable for verifying the functional correctness of the written HDL code. This approach is graphically illustrated in Figure 5-1.

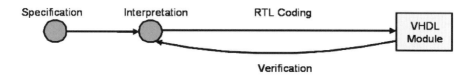

Figure 5-1. Interpretation from Specification to RTL

With the advent of TLM, the SoC design flow has significantly altered. If the TLM is coded independently of the RTL design by the same engineer, the corresponding flow is depicted in Figure 5-2. This figure shows clearly that the equivalence of the TLM and RTL design, i.e. *reconciliation*, must be proven. If the TLM design is fully tested as a golden reference model, and the reconciliation between TLM and RTL is proven, then the functional verification of the RTL design can be claimed proven. Note that engineers are once again held accountable for verifying the functional correctness of the written TLM codes.

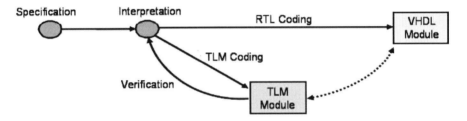

Figure 5-2. TLM-RTL Reconciliation

A better alternative of applying TLM in the SoC verification flow is the *redundancy* approach. This is a method where the RTL and TLM designs are conceived by different engineers. As illustrated in Figure 5-3, hardware engineers are in charge of the RTL design while system engineers are responsible for the TLM design. The term "redundancy" refers to the dual interpretation of the same specification. These interpretations, however, are deduced by two design teams of distinct approaches. The advantage of such difference in the interpretation can nicely catch and fill up the behavioral discrepancy between the RTL and TLM models.

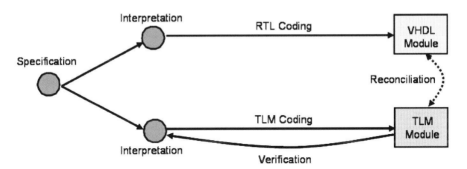

Figure 5-3. Redundancy Approach

Figure 5-4 represents another view of the verification flow, which is equivalent to the case shown in Figure 5-2. The difference is that the conversion from TLM to RTL is automated by a tool such as *behavioral compiler*.

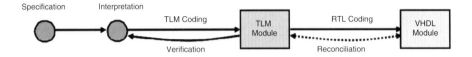

Figure 5-4. TLM to RTL Automation

There exist other structural organizations of the SoC verification flow using the TLM methodology. Be it any organization, the functional equivalence between the RTL and TLM models must always be respected. The functional verification of the RTL DUT is only valid with the conditions that such equivalence is observed and the TLM model is proven as a golden reference.

Hardware designers have a common fallacy of assuming the TLM-related activities as the unnecessary work of little value. This is unquestionably a false impression. The TLM methodology has many advantages to offer to its users (see Chapter 2); among which, reducing the workload of verification engineers is the advantage most pertaining to the functional verification.

Since it is indeed the executable specification of a given design, TLM can replace the manual process performed by verification engineers to generate the expected results of test scenarios. The generated result serves as the golden reference for the functional verification of the given design. This really saves the verification engineers a significant amount of time.

After all, a golden reference is always necessary in the test benches intended for validating the functional behavior of an RTL design. This reference model can be a TLM model that requires very little extra effort to build up compared to the conventional C or e models, *but* offers high reusability in the software development and the architecture analysis.

4.   PRINCIPLES OF APPLYING TLM IN SOC VERIFICATION FLOW

As seen earlier, introducing TLM can influence the SoC design flow. Many use models of the TLM verification exist. This section describes the principles of applying TLM in SoC verification flow. Three orthogonal axes are identified to decompose the constellation of the TLM use models in SoC verification.

The first axis expresses the difference of the availability between TLM and RTL models. Although the TLM model can play the role of the golden model in most of the use models, the RTL model will sometimes be the one. The second axis focuses on what should be compared between the two models to ensure their "reconciliation". Here, the DUT can be considered as a black-box or a white-box. The third axis is related to how the comparison is done between the two models; it can be performed during the simulation or at the end of the simulation.

4.1   First Axis: Which Model is the Golden Model?

Most of the time, the TLM model plays the role of the golden model. This is the case when the TLM model is developed at the very beginning of the design flow. It is indeed the best case because it allows developing and debugging the software at the beginning of the flow. In addition, the TLM development is much faster and the debugging is much easier.

In certain cases, the RTL model is ready before the TLM development. As an example, this situation is encountered when dealing with the backward compatibility. The RTL model has been developed, tested, and perhaps, has already been fabricated on an existing chip. The development of the TLM model under this situation is normally decided for the future development of the same IP but with additional features.

4.2   Second Axis: States or Transactions Checking?

The focus here is on what should be compared between TLM and RTL models. Indeed, this is very close to decide if the DUT should be considered as either a white-box or a black-box.

The black-box use models do not explicitly make use of the information held by the internal structure. Black-box tests usually focus on testing functional requirements. On the other hand, the white-box use models allow peeking inside the "box". They concentrate particularly on using the internal information of the IP to guide the selection of the test data.

It is absolutely coherent to require an exact matching of the traffic occurring on the bus interfaces of both models, which implies at least a black-box approach. The transactions occurring during the simulation must match exactly between the two models. It is therefore important to choose the right level of abstraction for the transactions.

Regarding the internal structure of the TLM with respect to the RTL, the requirement is open. A good practice is getting a TLM model with the same register structure as the RTL model. In such cases, it is possible to take it into account for a white-box approach as it splits the complexity of the model and facilitates the debugging.

4.3   Third Axis: Methods of Comparison

The ultimate goal of the TLM verification methodology is to verify that the output of both RTL and TLM models are *functionally* the same. As discussed earlier, the "golden reference" of a given DUT must be obtained by simulating its TLM model in a TLM test bench, followed by simulating its RTL model in the same TLM test bench for its functional verification. The simulated results of the RTL DUT are subsequently compared to the "golden reference". There are three methods for comparing the results:

- *End-Simulation Comparison.*
  At the end of both TLM and RTL simulations, the last states of the memories and registers are dumped into a file for comparison.

- *Simulation-Parallel Comparison.*
  An approach to conduct the comparison during the simulation process. Both TLM and RTL simulations are run in parallel. A software will extract the relevant information from both simulations for comparing their memory, register content, and simulation events. This approach provides a much finer verification grain at the expense of slower simulation time in a test bench environment that is more complex. Such expense, however, is compensated if the test meets a bug and fails. The simulation is stopped whenever the test encounters a bug. It means that the simulation does not have to keep on until the end of the entire test. As a result, the simulation time for verification is actually shorter and the debugging becomes easier. The main difficulty for this method is to find a good sampling instant of the data. It is not always possible to catch such sampling instants despite the freedom of choosing the abstraction level for transactions. Nevertheless, this approach prevents the divergence of the data order when integrating the TLM model at the top level.

- *Scenario-Embedded Comparison.*
  A self-checking test that embeds the comparing mechanism in the test scenario. The golden reference is included in the scenario so that the comparison of simulation results can be launched directly by the scenario during the simulation.

4.4   The Use Model in STMicroelectronics

Section 6 explains the strategy developed in STMicroelectronics for the TLM verification. It is a strategy positioned globally in the SoC design flow. The TLM model is the first model to be developed in this strategy. It will play the role of the golden model to perform the verification of the RTL model. The comparison of results is conducted mainly on those transactions occurring on the buses. However, the strategy is equipped with the capability of accessing the interval structure, similar to the one in "gray-box" approach. At the end of the simulation, a checking will be performed.

In Section 7, a TLM use model is illustrated. The golden model is the RTL and the checking is performed on-the-fly with intrusion in the DUT.

5.   ASPECTS OF TLM VERIFICATION FLOW

5.1   Test Bench

The test bench environment of the TLM-based functional verification is a TLM platform written in a system level language, for instance, SystemC. As a common practice, the preliminary step is to instantiate the TLM model of a given DUT within a TLM platform so as to develop the corresponding reference model. Once the reference is attained, the RTL DUT will replace the TLM DUT in the test bench to go through the functional verification.

The idea of proving the equivalence between RTL and TLM models is rather simple: use the same test bench to examine both models and then verify whether their output values are the same, as pictured in Figure 5-5.

Any discrepancy observed in the output implies that the two models behave differently. If an error is uncovered, it could be a misinterpretation of the design specification in one or both of the models. The debugging process can be started right after comparing the output values of both models. Although chances to find errors in the golden model are less than finding them in the DUT, an error could come from any of the two models if a discrepancy is observed. Since the TLM model works at higher level of abstraction, they are easier to develop and thus contain less bugs compared to the corresponding RTL model. For this reason, it is natural to use the TLM model as the golden reference model.

The approach explained above is a module-level test although the test bench runs a top-level simulation. The functionality of each module is tested without any interactions with other modules. Such test benches are categorized as the IP test bench. Another category of the test bench is the system test bench, which tests the behavior of a given IP integrated into a system as SoC. The remainder of the section further describes both test bench categories pertaining to our development work[1].

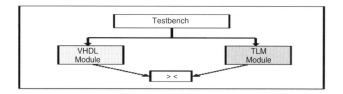

Figure 5-5. TLM Test Bench

5.1.1   IP Test Bench

The IP test bench is an environment to test a given DUT as an isolated component. Figure 5-6 illustrates the basic IP test bench that contains a test bench master, a memory, and a channel as described below:

- *Test Bench Master.*
  A test bench master is the initiator module that executes the test scenario. It could either be an ISS of a processor core intended for cross-compilation or a SystemC TLM module intended for native compilation. A particular SystemC component, *verif_host,* is designated as the test bench master for our in-house TLM-based functional verification.

- *Memory.*
  This is the data manager of an IP test bench. It loads the initial input stimuli associated to a test scenario. In addition, it stores the intermediate and final output generated during the test execution. A particular SystemC component, *verif_memory,* is designated as the test bench memory for our in-house TLM-based functional verification.

- *Channel.*
  A channel serves as the transaction router in an IP test bench. All of the components on a given test bench must be interconnected by a channel through which transactions are routed about for data exchanges. Our in-house TLM-based functional verification provides users with two choices of the channel. First, a simple route based on the TAC protocol. Second, a TAC channel with the log file instrumentation, *verif_channel.*

---

[1] Discussion is based on our in-house transaction accurate communication (TAC) protocol. Other protocols can be developed for TLM verification following the same methodology.

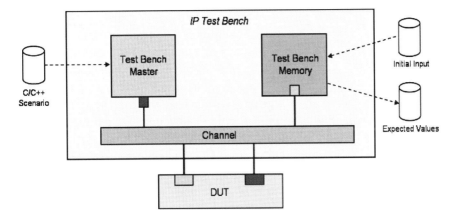

Figure 5-6. IP Test Bench

5.1.2   System Test Bench

A system test bench is an environment that tests the behavior of a given DUT integrated into a system as SoC. It could simply be an extension of an IP test bench in which other TLM IP models are added to form a system environment. Figure 5-7 illustrates such a test bench.

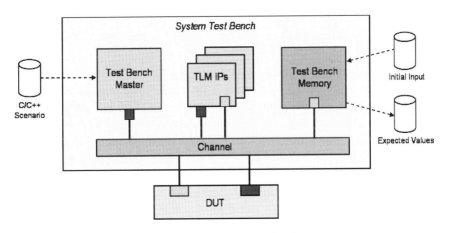

Figure 5-7. System Test Bench

5.2   Tool-Independent Aspect

The TLM verification is a tool-independent strategy, meaning that it is a methodology conceived in such a way that it can be applied for a wide range of tools. Although each particular tool has its own specificities, a number of common features are vital in order to apply the TLM verification strategy.

First, the RTL-TLM co-simulation is a recommended feature to have in the target tool. Most of the simulators available in the market today are able to co-simulate Verilog, VHDL, and SystemC models. It is still possible to apply the TLM verification strategy with a tool lacking the co-simulation feature. The verification task will nevertheless become more complicated.

Second, the target tool should offer an API for probing and controlling the RTL signals at any hierarchical level as well as for accessing to the TLM models. The probing and controlling of RTL signals exist in Verilog but not VHDL. Nowadays, most of the EDA tools offer a solution for this. Quoted below are a few examples:

1. Cadence has extended the SystemC classes of *sc_signal* using the functions of *observe_foreign_signal* and *control_foreign_signal*.
2. Modelsim has added particular features to "spy on" the monitoring, driving, forcing, and releasing of RTL signals.
3. Specman has indeed installed this feature since long. It has recently improved it with the introduction of ports (version 4.3.1) that allows connecting an e unit to another entity.
4. OpenVera uses *Virtual Port* for the connection of such purposes.

5.3   Characteristic of TLM Model

5.3.1   Abstraction Level

In the TLM-based verification strategy, TLM and RTL are two models required for a given DUT. Being two distinct modeling strategies, TLM and RTL models have different levels of abstraction. TLM DUTs are untimed transaction-accurate models for developing the reference simulation results while RTL DUTs are timing-accurate models for undertaking the functional verification. Due to such differences, the input and output of both models are consequently of different abstraction levels as well.

The difference in the abstraction level directly influences the way that different models are connected to buses. For RTL models, *input and output signals* are connected to buses. For TLM models, on the other hand, *read and write accesses* are connected to buses. It is therefore necessary to convert the accessing nature from one abstraction level to another.

For this reason, specific bus adaptors called bus functional models (BFM) are required to convert transactions into signals and vice versa between TLM test benches and the RTL models (see Section 5.4). Some connections such as interrupts stay at the signal level for both TLM and RTL models. Others, such as clock and reset, only exist at the RTL level.

5.3.2   Model Timing

As discussed in Chapter 2, there are two types of TLM models. First, untimed TLM models that describe the architecture of a given design. Second, timed TLM models that cover details of a given design down to the timing information at the micro-architectural level.

Although both models can be applied in the TLM-based verification strategy, untimed models are much more suited because they only capture the system specification, independently of implementation details. Indeed, the functional verification aims at verifying that the system behavior can hold regardless of the implementation choices.

If a timed TLM model is used as the golden reference, the objective will be different. It addresses the question: "Is the RTL implementation in compliance with the micro-architecture captured in the timed TLM model?" This book will cover mainly the untimed approach. Comments will be given for the timed approach.

5.4   Conversion of Signal and Transaction

In the conventional RTL verification, the RTL DUT is simulated in an RTL test bench. The DUT and the test bench communicate through a single sort of communication media: *signal*. In the TLM functional verification, on the other hand, the RTL DUT is simulated in a TLM test bench. The RTL DUT communicates with other modules on the TLM test bench through cycle accurate *signals* whereas the TLM test bench communicates with other modules through non-timed *transactions*.

Note that RTL and TLM are two distinct models of different abstraction levels. Their difference is particularly noticeable in their interfaces with the external world.

Table 5-1 indicates that TLM models have four types of interface while RTL models only have two. Be it a TLM or RTL design, these interfaces are specific to the bus protocol used in the design. The data flowing through the RTL and TLM models are of different formats as well due to such differences in interface.

Table 5-1. TLM vs RTL Interfaces

TLM Interface	RTL Interface
Input signal	Input signal
TLM Interface	RTL Interface
Output signal	Output signal
Master/initiator interface	Bus interface : Bundle of Input and Output signals
Slave/target interface	Bus interface : Bundle of Input and Output signals

Therefore, getting the two worlds to talk to each other necessitates some adaptations for converting signals into transactions, and vice-versa. This is feasible through implementing special adaptors, including pin convertors, signal convertors, bus functional models (BFM), and monitors. Figure 5-8 illustrates the global view of such adaptations.

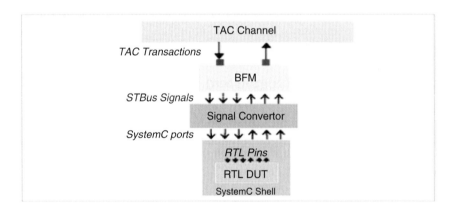

Figure 5-8. Global View of Signal-Transaction Conversion

5.4.1   Pin Convertor

The RTL DUT has a set of RTL pins to communicate with the external world. The cycle accurate signals are sent or received through these RTL pins. In the TLM verification strategy, the first step of the signal-transaction conversion is to create an equivalent set of these RTL pins in a format compatible with the TLM modeling.

Since we have chosen SystemC as our modeling vector, these pins are naturally to be converted into SystemC pins. A pre-requisite for doing so is implementing the STBus interface in the RTL DUT to allow applying the relevant BFM in the TLM verification.

A SystemC shell, *sc_foreign_module²*, is developed around the RTL DUT to convert the RTL pins into the equivalent SystemC ports. Tools are available to perform this pin conversion automatically from the top module of the RTL DUT. Some tool examples include *scv_shell* proposed by our in-house TLM verification strategy and *ncshell* proposed by NC-Sim.

Throughout the TLM verification, the communication between the RTL and SystemC ports are handled by the co-simulation layer of the NC-SystemC simulation kernel.

By default, the SystemC ports are converted to the signal type closest to the corresponding RTL signal type. This matching is normally performed by the pin convertor tool in the signal type optimization. Figure 5-9 describes the idea of pin conversion.

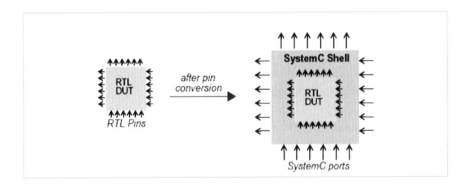

Figure 5-9. Pin Conversion

5.4.2 Signal Convertor

Once the RTL pins of a given DUT are converted into TLM ports (i.e. SystemC ports for our case), the subsequent step is to implement a signal convertor required for accommodating the following needs:

- *Signal Type Conversion.*
  Recall from Figure 5-8 that there is a BFM layer connected to the TLM channel of the test bench. The converted TLM ports do not always have the same signal types as those of the BFM ports. A proper adaptation of the IP-specific signal type by a signal convertor is therefore necessary.

- *BFM Signal Buffering.*

---

² If NC-Sim is used, the corresponding shell will be *ncsc_foreign_module*.

The initiator ports and the target ports have different bus protocol rules. Connecting these ports directly to the converted TLM ports give the same signal waveform for both intiator and target cases, which violates seriously the dissymmetrical protocol of STBus. To avoid this problem, bus request and bus grant signals must be buffered by a signal convertor to guarantee a flawless signal synchronization.

- *Timing Constraints Insertion.*
  Timing constraints need to be inserted by a signal convertor for adding random delays in the signal propagation between the BFM and the converted TLM ports. Such timing constraints stress the IP under test, and consequently fill up its internal FIFO. These random delay values are chosen and entered by the users in an *xml* configuration file.

- *Cycle Delay Insertion.*
  This is a constant quarter-periodic cycle delay to prevent the ambiguity in the delta cycle of the verification. Such ambiguity or vagueness may occur if the synchronization signals happen to change at the same time as the bus clock signals.

5.4.3   Bus Functional Model

A protocol is a set of rules imposed between an initiator and a target in a given system. Initiators and targets drive signals of different purposes. The typical signals driven by initiators are those with request, address, or data information; while the most common signal driven by targets is the acknowledge signal. Both initiator and target modules must drive their signals by respecting the protocol rules.

The TLM strategy does not intend to drive these signals directly. Instead, a bus functional model (BFM) is inserted between RTL and TLM models. The BFM is indeed the protocol adaptor for a transaction level bus model. It converts RTL signals of the RTL DUT into TLM transactions for the TLM test bench, and vice versa. The following discussion on BFM refers to the STBus BFM, which has been developed for our in-house TLM verification based on the STBus[3] protocol.

The discussion can easily be generalized to other modern bus protocols such as AXI[4] and OCP[5] because they share the same fundamental concepts, such as asynchronous request and response to enable queuing of transactions and out-of-order responses.

---

[3]   Refer to http://www.stmcu.com/inchtml-pages-STBus_intro.html
[4]   Refer to http://www.arm.com/products/solutions/AMBAHomePage.html
[5]   Refer to http://www.ocpip.org

The STBus BFM can be implemented either for an initiator port or a target port. Three types of STBus protocol, T1, T2, and T3, determine the implementation choice of an STBus BFM. If the RTL interface is an initiator, i.e. an initiator port of the RTL DUT needs to be connected to a TLM target, then an *initiator BFM* must be implemented to drive signals from the RTL initiator to the TLM test bench. On the other hand, if the RTL interface is a target, i.e. a target port of the RTL DUT needs to be connected to the TLM initiator, then a *target BFM* must be implemented to drive transactions from the TLM test bench to the RTL DUT.

Figure 5-10 clearly illustrates the structure of both initiator and target STBus BFM. Note that each BFM is divided into two adaptors, i.e. cell-packet adaptor and packet-transaction adaptor. To handle the signal-transaction conversion, the BFM is split in three different levels:

1. Block level.
2. Packet level.
3. Cell level.

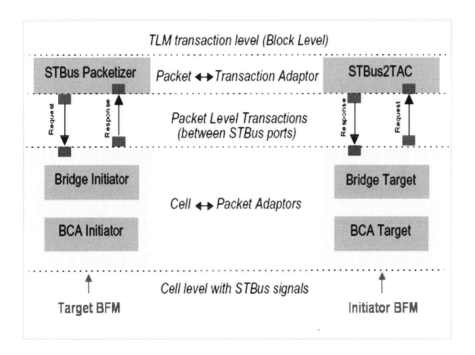

Figure 5-10. Bus Functional Model

Consider the case of the initiator BFM. The STBus signals from the RTL DUT are observed at the cell level. These signals are passed to the cell-packet adaptor and then converted into the packet level transactions. A bus cycle accurate (BCA) element and a bridge are used in the cell-packet adaptor. Once converted, the packet level transactions are transported through the STBus ports into the packet-transaction adaptor, i.e. the STBus2TAC. This second adaptor converts the packet level transactions into the block level TLM transactions, which are subsequently connected to the TLM test bench via the TAC ports.

The same idea is applied to the target BFM by following another sense of conversion, starting from the TLM transaction at the block level to the STBus signal at the cell level.

In a given design, users normally need a BFM to go from the cell level to the block level (C2B), and another to go from the block level to the cell level (B2C). Since there are three types of STBus protocol (i.e. T1, T2, and T3), six different combinations are possible: C2BT1, C2BT2, C2BT3, B2CT1, B2CT2, and B2CT3. Each of these combinations can be instantiated as a single BFM component as depicted in Figure 5-11.

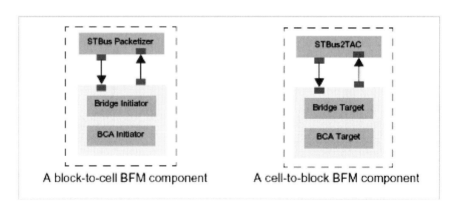

Figure 5-11. Encapsulated BFM Component

A clock and a reset component are necessary for handling the cycle accurate part in the TLM verification. Several clocks are configurable dynamically in an RTL design. A bus clock of the same frequency as the RTL clock must be implemented in the BFM. A particular TLM reset component, *verif_reset*, is developed to configure the form and the duration of the reset signal in the BFM.

5.4.4 Monitor

Monitors are used in some use models, for instance, for the on-the-fly comparison of TLM and RTL models in an RTL test bench.

As explained earlier, a BFM is implemented in the TLM verification to set up the communication between:

1. an RTL initiator interface and a TLM target interface;
2. an RTL target interface and a TLM initiator interface.

In the TLM test bench, however, no communication occurs directly between RTL and TLM modules. Since the RTL interfaces are connected to the test bench through the BFM, the communication occurs indeed within the TLM test bench. Therefore, an important aspect in the TLM verification is to *monitor* the activities taken place in the RTL models. Two major activities to be observed and supervised are:

1. transactions sent by an RTL initiator interface must be compared to those sent by the corresponding TLM initiator interface;
2. transactions transmitted to an RTL target interface must also be sent to the corresponding TLM target interface.

To carry out such checking, no signals need to be driven. Instead, signals are *probed* in RTL models to detect any transaction sent by an initiator interface to the TLM test bench as well as any transaction received by a target interface from the TLM test bench.

No BFM but a *monitor* is required to perform such checking task. The monitor is place between the RTL and TLM designs. Similar to the idea of BFM, there are initiators and target monitors depending on the nature of the connected RTL interfaces. Figure 5-12 demonstrates the difference between the monitor and BFM.

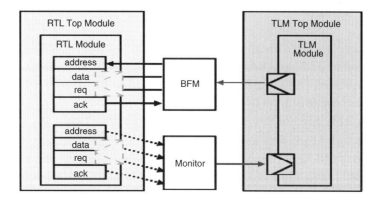

Figure 5-12. Difference between BFM and Monitor

5.5   Analysis of Simulation Results

5.5.1   Output Data for Comparison

Three categories of output can be classified for the TLM model:

1. *Initiator Interface Output.*
   For an initiator interface, the output is the entire transaction, i.e. address and data information.

2. *Target Interface Output.*
   For a target interface, the output is the transaction status such as success, failure, waiting, etc. This is typically a return data value for a read access, for instance, a successful read access will have the transaction status of "success" as the output.

3. *Signal Output.*
   TLM models can also have signals as output.

5.5.2   Implementation of Checker

In the TLM verification, the output data sampled in the RTL and TLM models are compared and verified if they meet the following conditions:

1. The output values are the same.
2. The outputs are in the same order.

These conditions must be fulfilled to guarantee an equivalent functional behavior between the RTL and TLM designs.

The checker has two FIFOs: one dedicated for collecting RTL values and another for collecting TLM values. The matching values are stored in the appropriate FIFO when they are available. As soon as a transaction is completed on both the RTL and TLM sides, stored values are retrieved from the FIFOs for comparison. The length of the checker FIFOs depends largely on the design.

5.5.3 Visualization of Results

To help users better interpret the result of a given functional verification, the TLM verification methodology should allow the integration of useful tools to visualize the simulation results.

Regarding this aspect, our in-house TLM verification is designed to support the Cadence NC-Sim logic simulator in order to provide users with a graphical interface for signal observation and control. A handy monitoring mechanism is made available by integrating another in-house tool, *SysProbe*.

Furthermore, a score board mechanism is employed for relating bus requests to bus responses through SysProble. It also offers the custom analysis such as tracing the divergent point between RTL and TLM simulations.

Chapter 6 gives further discussions on the performance analysis and verification, including the SysProbe tool.

6. STRATEGY OF TLM VERIFICATION

The functional verification based upon the TLM methodology employs a strategy consisting of four phases:
1. development of test scenarios and the associated input stimuli;
2. execution of test scenarios on the SystemC DUT in a TLM test bench to acquire a golden reference;
3. execution of test scenarios on the RTL DUT in the same TLM test bench to conduct the functional verification of the DUT;
4. analyses of test results by comparing output data and expected values.

A verification test is composed of a test scenario and its associated input stimuli. Two test categories can be distinguished according to the test nature:

- *Architecture Test.*
  It tests the specific behavior of a given DUT regardless of its timing implementation. The behavioral aspects covered are principally the DUT configuration and control, meaning that some tests are initialized and executed by the architecture test scenario on the DUT.

- *Micro-architecture Test.*
  This is a test created by incorporating the timing behavior of a given
  DUT into its architecture test. Such timing behavior is implemented at
  the cycle level to observe the timing impact on the DUT.

6.1   Test Scenario

This section discusses test scenarios along with some implementation
details relevant to our own development work.

6.1.1   What is a scenario?

In a more common term, scenario means test code. Test scenarios applied
in the TLM verification strategy could be test programs written in C or C++.
Indeed, the test scenario plays a very important role in the verification flow
because the quality of a verification test depends largely on the scenario.

A scenario initializes, configures, programs a given DUT and ensures the
execution of a particular control sequence. The same scenario could be
employed for an architecture or micro-architecture test.

To ensure a good test coverage, the test code of a scenario comprises two
kinds of test: directive test and random test (or a mixture of both). A
directive test is fixed by test developers whereas a random test is arbitrarily
generated. TLM is able to link the SystemVerilog[6], e and SCV[7] random
generators in order to allow test developers writing a script in e language
that generates the random test of a scenario.

6.1.2   How does a scenario work?

Since a scenario is applied in a TLM test bench, the information about
the address mapping must be provided. We handle this by providing the
necessary information in two address mapping files that are included by the
test code of a given scenario:

- *mapping.h*
  Holds the overall system address mapping information of the test bench.
- *IP_mapping.h*
  Provides address mapping information specifically related to an IP under
  test, i.e. DUT.

---

[6]   Refer to http://www.systemverilog.org/
[7]   Refer to http://www.testbuilder.net/

The test scenario utilizes a particular DUT API, *bus driver*, to write into or read from a DUT. Two principal functions are implemented through the *bus driver* API:

- *DUT_write ()*
  This function enables the test scenario to write into a DUT by calling the standard *master_port.write()* in the host ports.
- *DUT_read ()*
  This function enables the test scenario to read from a DUT by calling the standard *master_port.read()* in the host ports.

Note that a set of *DUT_write()* and *DUT_read()* primitives could be assembled to build higher level IP programming functions. Once a test scenario is written, it is compiled in a Dynamic Link Library (DLL) to yield a *.so* file. The resulted *.so* file is ready to be dynamically linked to the TLM test bench for its execution. Figure 5-13 sums up our discussion graphically.

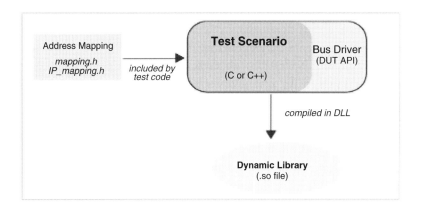

Figure 5-13. Working Principle of Test Scenario

6.1.3   Where does a scenario go?

Once compiled, a test scenario is dynamically linked to the simulation kernel for its execution. It is executed by the host of a TLM test bench, which could either be an ISS of a processor core for cross-compilation or a SystemC TLM module for native compilation.

We have developed a particular SystemC TLM component, *verif_host,* as the test bench master. Indeed, it is a generic native wrapper called *Runner DLL*. This wrapper contains an internal SystemC thread, which calls the main function of a test scenario in order to execute the test program.

6.2   Input Data

Input data is the initial value loaded into the test bench in a functional verification. It is indeed a set of input stimuli associated with a given test scenario. These input stimuli are dependent on the IP under verification.

Very often, a given group of input data forms a test set specifically suited for testing the functions of an IP. For instance, an MPEG IP usually has a specific test set to decode the image. The role of input data becomes crucial if it influences the behavior of the DUT. In general, there are two types of input data:

*   *Bitstream*. Data set for performing tasks as text and image processing.
*   *Firmware*. Data set required for enabling the DUT to execute a program.

The input data of a given test scenario is loaded into the data manager of the test bench. The data manager is generally a memory module as we shall discuss in section 6.4.

6.3   Expected Value

Once the input data is loaded into the test bench, the associated test scenario is executed together with these initial values. A reference model must be developed as the first approach to the TLM verification, hence the execution of the test scenario on the TLM DUT.

The resulted output of this execution is compared to the "initial golden reference" of the input data. Typically, the initial golden reference is computed by an algorithm developed earlier by the author of the design specification. Such comparison could simply be a manual interpretation of the design specification or the validation of a particular test set. If there is no difference between the resulted output and the initial golden reference, the resulted output can then be qualified as the *expected value* for the given set of input data.

The random generator can generate a large set of input data to enlarge the test coverage. This feature is critical if the input data could influence the behavior of the DUT. The randomly generated input data can help producing additional expected values.

The expected values are collected as the "golden reference" of the TLM functional verification. The TLM DUT can now be replaced by the RTL DUT in order to perform the functional verification. It means that the same test scenario is executed on the RTL DUT with the same set of input data. The resulted output of this second execution is compared to the "golden reference", i.e. the expected values obtained earlier. If any difference is detected between them, the trouble-shooting should be conducted for both TLM and RTL models through examining the design specification

Bear in mind that the probability of encountering failures in both TLM and RTL simulations with the same error occurrence is relatively low since TLM and RTL are very different methodologies.

6.4   Data Manager

The data manager is vital to manage the input data and the resulted output of a verification test bench. We have developed a generic SystemC TLM memory module, *verif_memory*, to handle this task.

The *verif_memory* component is indeed an instantiated C array accessible through a TAC slave port. It loads the initial data and stores the intermediate as well as final output generated during the test execution. This part is done within the memory table. The TAC slave port is dedicated to receive transactions from the TAC channel in the test bench.

There are three portable functions in *verif_memory* for data management and checking as listed below:

- *Loader*. The loader is an IP-dependent function. It selects and loads the initial data into the memory. The data loaded could be of binary or hexadecimal values. The loader function is replaceable without changing the memory model.

- *Dumper*. A dumper is a non-IP-dependent function. Its actions are controlled by the test scenario. The dumper has virtual registers in which the start and end addresses of a memory area are written for dumping the resulted output of a simulation.

- *Checker*. There are two standard checkers. First checker verifies that the non-initialized memory cell is not accessible. Second checker forbids writing twice the same value in the same location so as to ensure an optimal IP behavior performance. Both checkers could be enabled or disabled dynamically without re-compilation. It is quite easy to develop a purpose-specific checker since it is written in SystemC.

6.5   Global Test Environment

The platform of a test bench is composed of both RTL and TLM models. It is recommended to have two separate top modules, i.e. one for RTL and another for TLM. Although such mixed design is hierarchical, separating the top modules avoids altering RTL models intended for the synthesis as well as TLM models intended for the software simulation.

Such approach is allowed by certain EDA tools, for instance, the Logic Design Verification (LDV) tool of Cadence as illustrated in Figure 5-14. For those tools that do not support the configuration of separating top modules, two sub top modules should be prepared for each model as depicted in Figure 5-15. As an example, the ModelSim of Mentor Graphics only permits a single top module in RTL.

Figure 5-14. Separate Top Modules as Test Bench Platform

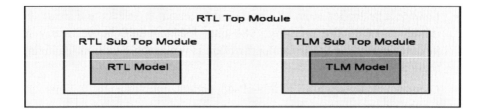

Figure 5-15. Single Top Module as Test Bench Platform

The global test environment of the TLM verification test platform is demonstrated in Figure 5-16.

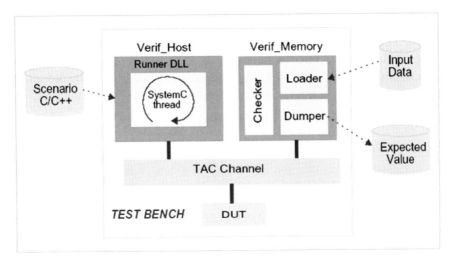

Figure 5-16. Test Environment of TLM Verification

6.6 Automated Tool

We have developed tools based on the SPIRIT standard to automatically generate a TLM verification test bench with the necessary adaptations for the signal-transaction conversion. See Chapter 7 of this book for further descriptions on the subject of design automation.

7. MANAGING LEGACY TEST BENCHES

The methodology of TLM verification has succeeded its introduction phase smoothly. Yet, it may not be obvious for designers to invest time in writing up TLM models of their design.

Many designs are developed from some existing work. Indeed, a "new" RTL design is quite often an adaptation of the RTL work done previously. Such situation raises frequently the legacy issue. This issue, however, can be alleviated by the TLM verification methodology.

In general, an existing RTL design has already been intensively tested, synthesized, and routed. Through the TLM verification methodology, this design can be handily reused as the golden reference to develop and test the equivalent TLM design. Once accomplished, the TLM design will be a good starting point to add new features. Such approach unquestionably helps to consolidate the introduction of TLM as a novel verification methodology.

This section provides the test cases of two real implementation examples for the TLM verification. The goal of each implementation is to build the TLM models of the next generation IPs. The first example is an ADSL modem design based upon an RTL test bench in VHDL, whereas the second example is a Video Encoder/Decoder design using an RTL test bench in e language.

7.1   ADSL Modem Test Case

7.1.1   Test Environment

The need for the Samx ADSL modem test case raised from the software developers because a platform was required to test the software design as soon as possible in the design flow.

A test environment targeting to detect 80% of the bugs was therefore set up for the modem. The characteristics of such test environment were:
1. The modem design was in VHDL.
2. Module and top-level test benches were in VHDL.
3. The simulator was LDV 5.0-s013 from Cadence.
4. The TLM design was written in SystemC 2.0.1.

7.1.2   Test Bench

The test bench of the Samx ADSL modem is described in Figure 5-17.

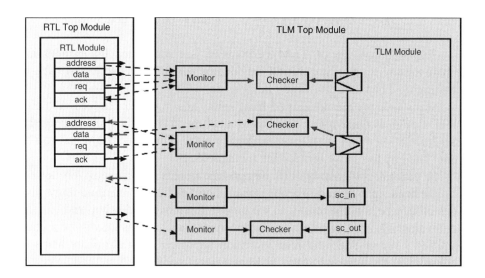

Figure 5-17. Test Bench of Samx ADSL Modem

All of the monitors and checkers in the test bench were developed in SystemC. The communications from SystemC modules to RTL modules were established by the SystemC library extension of Cadence. The SystemC *sc_signal* could be connected to RTL signals through the functions of "*observe_foreign_signal*" and "*control_foreign_signal*". The monitors would reflect the values of the RTL interfaces, which were sent to a checker or to the RTL module depending on the fact if they were output or input of the module.

7.2   Video Encoder and Decoder Test Case

7.2.1   Test Environment

The need for the video encoder/decoder test case raised from the hardware designers for two reasons:
1. Get ready a TLM platform to develop the next generation design.
2. Alleviate the use of the e test bench by reusing TLM modules.

The test case was intended for testing a particular video encoder/decoder, *pxp_hamac*. A Specman verification environment of Verisity was set up to test the RTL model of this module. The "Verification Advisor" methodology proposed by Verisity employed the concept of golden model. The golden model was normally developed in e or C language. The principle was the same as TLM verification, i.e. the same input data were applied to both golden model and DUT followed by a comparison of their output values.

The initial golden model for *pxp_hamac* was developed in C language. The TLM models were conceived based on the C functions. These C functions were encapsulated by a SystemC wrapper containing:
*   interfaces to the external world;
*   synchronization processes to synchronize inputs of modules.

Advantages of replacing C functions by TLM models in developing the golden model of the test bench included:
*   testing a real model to be integrated in a simulation platform;
*   significant simplification of the e test bench.

The characteristics of the test environment were:
1. The design of the video encoder/decoder was in Verilog and VHDL.
2. The module-level test bench was in e language.
3. The simulator was LDV 5.0-s013 from Cadence.
4. The TLM design was written in SystemC 2.0.1.
5. The version of Specman was 4.3.3.

7.2.2   Test Bench

The test bench of the *pxp_hamac* video encoder/decoder is described in Figure 5-18.

All of the monitors, checkers, and BFMs are developed in e language. The test stimuli were injected from the e environment. For master port accesses, stimuli were generated at the transaction level and sent to the RTL module through the BFM.

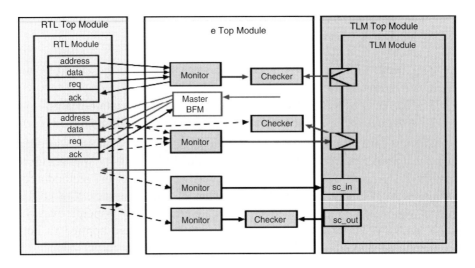

Figure 5-18. Test Bench of *pxp_hamac* Video Encoder/Decoder

7.3   Remarks

The two examples given above illustrate the different implementations based upon the same methodology.

The common points for both examples are:

- Both test bench platforms tested SystemC TLM models and compared the results to those of the corresponding RTL models.
- The original test benches used for testing the RTL design were reused for the TLM verification.
- No modifications were necessary in the RTL design.
  Differences do exist between the two examples:
- Stimulus source for the ADSL modem was VHDL while it was e language for the video encoder/decoder.
- The IP modeling language for the ADSL modem was SystemC while it was e language for the video encoder/decoder.

8.   TRANSACTIONAL CO-EMULATION

8.1   Introduction to Hardware Emulation

Given the SoC complexity today, the traditional HDL simulation is inefficient to perform SoC verification because HDL simulators are not fast enough to handle the high numbers of clock cycles involved.

The hardware emulation aims at providing verification engineers with the prototype of a given SoC under verification. Such prototypes are based upon reconfigurable components, for instance, FPGAs or CPUs. The maximum visibility to the interior of the SoC design must be given by these prototypes. The ideal emulator is a system that enables designers to observe all signals with the nominal system frequency while running their SoCs. Unfortunately, such perfect emulators do not exist.

Depending on the emulation use models, the existing emulator solutions allow running verification at a frequency ranging from 1kHz to more than 10MHz (compared to 10Hz afforded by a typical simulator). These solutions provide users with some visibility on the design to help debugging. Such visibility ranges from some few signals up to a full observability (just as the visibility provided by simulators). Most of the emulation solutions are capable of showing latches, memories, and registers, which are the most important elements to be observed.

Several use models are available for emulating a system. Each use model is characterized by its own complexity level, frequency range, debugging possibilities, and time-to-emulation. Some of the most common emulation use models are collected herewith.

- *Self-Test Bench Emulation (STB)*
  In the STB mode, both synthesizable test bench and DUT are run by the emulator at a typical frequency of about 1MHz. The bottleneck of this mode is obviously the time-to-emulation because much time has to be invested in writing a synthesizable test bench. Furthermore, users must prepare a specific yet hard-to-reuse test bench.
- *In-Circuit Emulation (ICE)*
  The ICE mode runs a given DUT on an emulator while the emulator is linked to a real environment. To verify a video chip, for example, a video camera and a monitor can be plugged in to observe if the system demonstrates the expected behavior. This emulation mode is well adapted for developing the chip firmware. However, it generally requires high emulation frequencies due to the real environment constraints. This mode is not quite appropriate for hardware debugging because of two reasons:

a) emulators can hardly cover both high speed and high visibility;

b) ICE environment does not provide a full synchronization that ensures a deterministic behavior at cycle level.

- *Vector Debug (VD)*

  In the VD emulation mode, a given DUT runs on an emulator with the input stimuli and the expected outputs pre-computed for each clock cycle. Indeed, this use model is analogous to using the emulator as a production tester that provides an internal visibility to the chip.

- *Cycle Accurate Co-Emulation*

  The cycle accurate co-emulation runs a given DUT on an emulator while running the test bench on a workstation linked to that emulator. Interface signals are updated at each clock cycle. The distinctive attributes of this mode include low time-to-emulation, test bench reuse, and the typical frequency value of about several kHz.

- *Transactional Co-Emulation*

  The transactional co-emulation is quite similar to the cycle accurate co-emulation. The key difference is the way they handle the synchronization between software and hardware environments. Here, interface signals are exchanged only when a communication is required but not at each clock cycle. Next section will focus on discussing this use model.

Figure 5-19 compares the different emulation modes in terms of the clock frequency and time-to-emulation.

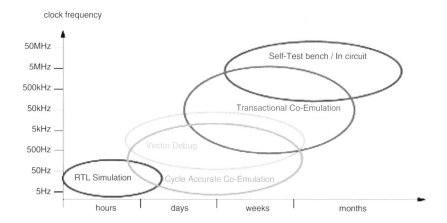

Figure 5-19. Common Emulation Use Models

8.2 Transactional Co-Emulation Mechanism

The transactional co-emulation runs a given DUT on an emulator within a software test bench. It is a high-level verification based upon actions instead of signal assignments. The working mechanism of such emulation is described hereafter.

In the transactional co-emulation, the software test bench generates some messages that are sent to the hardware. Referring to these messages, the hardware is able to work for several clock cycles. Likewise, the hardware may use several clock cycles to generate a message for the software.

The structure of the transactional co-emulation is depicted in Figure 5-20. Note that there are *transactors* in the dotted areas of the figure, which are specific hardware structures designated for receiving or creating messages. Essentially, transactors provide the same functionality as the BFMs discussed earlier. A transactor must contain at least one hardware port to send or receive transactions. Each of these hardware ports is linked to a software function named *proxy*.

Collected below are the benefits of using the transactional co-emulation:

1. Due to the transactional aspect, emulators can run close to the full speed such as in STB mode in the best case.
2. Due to the co-emulation aspect, fast and reusable C/C++ test benches can be easily developed.
3. Easy integration in a TLM/SystemC environment.
4. Verification can also be performed at the system level.

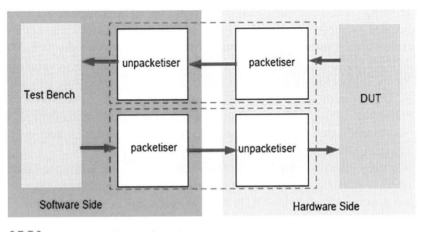

Figure 5-20. Mechanical Structure of Transactional Co-Emulation

Typical applications of transactional co-emulation are verifications of:
- stream data such as audio, video, and telecommunication data;
- memory access modeling;
- bus protocol modeling together with IP verification.

The major pitfall of the transactional co-emulation is the compulsory transactors that are not always easy to develop. A good investment to resolve this pitfall is to develop generic transactors. The libraries of the generic transactors may take quite long to build up but the tools based on the libraries then allow quick implementations.

8.3   Standard Co-Emulation Modeling Interface

The standard co-emulation modeling interface[8] (SCE-MI) is a message-passing environment that connects a hardware model to a software model. The SCE-MI standard is still under continuous development but the first release was successfully delivered in May 2003.

The objective of SCE-MI is to provide a free standard communication interface suitable for high performance transactional co-emulations. Through this standard, any transactors are able to run correctly with all the existing emulation solutions. Therefore, users will not waste their time any longer in developing specific transactors based on a proprietary emulator API targeting just a single emulator.

To increase its efficiency, the SCE-MI has defined two types of clocks:
- Uncontrolled clock - uclock
- Controlled clock - cclock

Note that the uclock is an *un*controlled clock, implying that this clock never stops. Transactors or BFMs work upon the rising edges of the uclock. The cclock is implemented in DUTs. In contrast to the uclock, the cclock is controllable by transactors. This interesting feature empowers the SCE-MI to send messages from test benches, which will be processed for several clock cycles by transactors. During a message processing, transactors can stop a DUT. Once the message is processed, transactors can feed the DUT with several clock cycles without stopping it anymore. This mechanism allows the use of dedicated algorithms for message coding, which in turn provide better speed efficiency.

Regarding the functionality, the SCE-MI provides four hardware macros:
1. *in port macro*, allowing DUTs to receive messages from the test bench;
2. *out port macro*, allowing DUTs to send messages to the test bench;
3. *clock macro*, generating DUTs clocks;
4. *clock control macro*, controlling DUTs clocks.

---

[8]   Refer to the web site of SCE-MI for further information: http://www.eda.org/itc.

In addition, the SCE-MI provides C++ functions (i.e. proxy) that can be linked to the in or out port of each hardware module. Figure 5-21 describes the detailed structure of a transactional co-emulation environment that employs the SCE-MI.

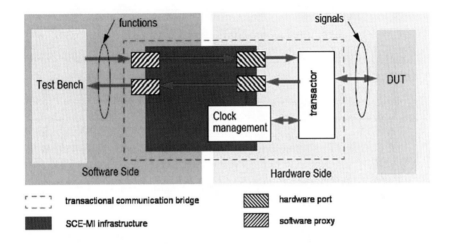

Figure 5-21. Applying SCE-MI in Transactional Co-Emulation

8.4   Applying TLM in Transactional Co-Emulation

The TLM is a transaction-oriented modeling seamlessly adapted to generate efficient test benches for the transactional co-emulation of a SoC design. When the TLM methodology is used in conjunction with the transactional co-emulation, designers can shorten the SoC design cycle as the same platform is employed ranging from high-level tests to verification.

SoCs comprise several IPs that communicate through a bus or network-on-chip (NoC). Through the TLM approach, the behaviors of these IPs are modeled and the SoC functionality can be simulated at an early stage.

In the transactional co-emulation setting, SoC verification tests can be performed by mixing TLM and synthesizable hardware IPs. Such concept allows the test bench and non-ready IPs run in the TLM simulation while the synthesizable HDL IPs run in the emulation mode. The same TLM test bench is consequently retained all through the SoC design cycle. Not only that an early checking of the SoC functionalities can be conducted, such methodology also enables high reusability of test benches and transactors. Figure 5-22 illustrates the principle of applying TLM in the transactional co-emulation.

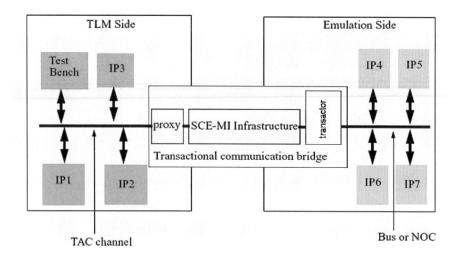

Figure 5-22. Applying TLM in Transactional Co-Emulation

8.5   Example

The methodology of combining TLM and the transactional co-emulation was successfully applied to the LCMPEG design[9].

As depicted in Figure 5-23, two different approaches to co-emulation were conducted for the LCMPEG design: cycle accurate and transactional co-emulations. The *co*-emulation environment of either approach was divided into two parts:

- *Software Part.* This was the TLM test bench written in SystemC. It comprised a host model (e.g. ISS) and a memory module.

- *Hardware Part.* The LCMPEG IP design was placed in the emulator as the hardware part.

---

[9]   Low Cost MPEG (LCMPEG) is an IP design fully developed and tested by STMicroelectronics France.

The LCMPEG design applied the STBus protocol as the communication interface. Essentially, this was the very part that drew the difference between the cycle accurate and transactional co-emulations.

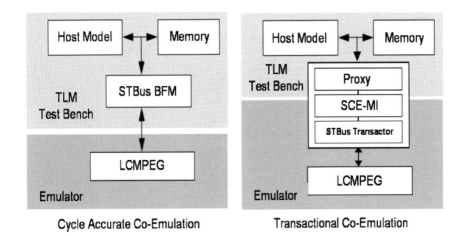

Figure 5-23. Cycle Accurate vs Transactional Co-Emulations

In the cycle accurate approach, the STBus BFM was conceived as a software element located in the software part. The "software" BFM functioned as a convertor to transform the TLM transactions from the TLM test bench into the corresponding signals to be sent to the LCMPEG IP in the emulator, and vice versa. The transaction-signal conversion was carried out entirely on the software side.

The STBus transactor for the transactional co-emulation, on the other hand, was designed as a synthesizable hardware element located in the hardware part. Indeed, the SCE-MI was employed in this approach as the "message passer" between the software and the hardware parts. The C++ proxy was implemented in the software test bench as the software entry point while the transactor was implemented in the hardware emulator as the hardware entry point. The proxy and the transactor communicated through the SCE-MI infrastructure. TLM transactions from the TLM test bench were transferred from the proxy via the SCE-MI to the transactor. Then, these transactions were converted into the corresponding signals by the transactor. The term, transactor, comes from the SCE-MI standard definition; but it is also called BFM as it performs the equivalent conversion as the BFM in SW. Note that the transaction-signal conversion was carried out entirely on the hardware part, meaning that the conversion was much faster (because implemented inside the emulator) compared to the cycle accurate co-emulation.

Several tests were conducted to emulate the LCMPEG IP. The design was first executed through the cycle accurate approach on an emulation tool at a maximum system frequency of about 17kHz. The "software" BFM was then adapted into the "hardware" STBus transactor. Through this adaptation, the transactional co-emulation was performed at the system frequency of 170kHz for the LCMPEG design. Note that this frequency was increased tenfold on the same emulator compared to the system frequency used in the first approach. The observation made on the cycle accurate co-emulation showed that the communication between the TLM test bench and the LCMPEG hardware DUT only occurred for 10% of the hardware clock cycles. Thus, 90% of synchronizations done between software and hardware sides were not required and they were actually slowing down the verification platform. Transactional co-emulation reduced this communication cost and hence improved the frequency of the verification platform.

8.6   Conclusion

To conclude, the LCMPEG example has demonstrated very well the benefits of using TLM together with the emulation platforms. Without emulation, a pure TLM co-simulation platform could only run at about 500Hz on the most powerful Sun workstation. A speed gain of about 30 can be achieved simply by replacing a simulator with an emulator. Furthermore, using the transactional co-emulation methodology gives a platform that is 300 times faster than the pure simulation platform.

Since TLM allows reusability, the time-to-emulation is getting shorter and shorter. Through such methodologies, designers can obtain a fast and efficient SoC verification platform in very short time-lag.

9.   FORMAL VERIFICATION

9.1   Introduction

The previous sections presented the TLM approach to verification through simulation. The methodology is nowadays applicable on large designs. The most common limitation of dynamic verification is that it cannot be exhaustive except for very particular situations.

Formal verification, on the other hand, can be exhaustive. Ideally, formal verification could prove the correctness of a model or uncover all the bugs it contains. The problem of formal verification, however, is the difficulty to scale up. The algorithms applied are usually exponential in the worst case, which makes the exhaustive and exact verification, although possible in theory, not always applicable in practice.

Many techniques have been developed to allow model checking to scale up. Most of them rely on approximations. To remain in a formal context, we have opted for the conservative approximations that add possible behavior but not removing any. The consequence of using such approximations is that the response from the verification tool is not *true property / false property* but *true property / do not know*.

Formal verification is therefore extremely interesting when applicable in real practices. It is nevertheless not an absolute substitute for verification by simulation because the scopes of both approaches are different.

This section introduces an approach exploiting all the particularities of a TLM design written in general SystemC through our tool implementation: LusSy.

9.1.1   Related Work

SystemC designs being *circuit* designs, we could thus consider using one of the verification tools (model-checkers, SAT-solvers, etc) developed for hardware verification such as SMV [1]. These tools, tailored for RTL designs, exhibit however a clear notion of logical time while we actually need to deal with heterogeneous designs. Heterogeneity comes from several areas: determinism and non-determinism, synchronous and asynchronous systems, hardware and software components. Moreover, these tools cannot accept general SystemC as inputs.

As far as we know, all of the work done to date on verification techniques and tools for SystemC designs are limited to the subset of SystemC that allows writing RTL designs. Such techniques or tools cannot be used for real TLM designs (refer to [2] for an example). Similar attempt made to treat SpecC language is extensively described in [3]. An execution date is associated with each instruction of the program. These dates can then be considered as a dependency graph, on which synchronization properties can be proved. The approach is very limited because only a single date can be associated to an instruction. Such limitation does not allow general loops to be taken into considerations.

Since SystemC is mainly a C++ library, one may expect confronting the same problems as those addressed by general-purpose software model checking tools. This is *not* the case. Verifying SystemC designs is, on the one hand *simpler*, because general dynamic data structures and general algorithms are not dealt with; on the other hand *harder*, because the parallelism and the scheduler specification must be taken into account.

General software model checking techniques concentrate on dynamic data structures and general algorithms. They provide sophisticated tools such as invariant extraction and loop unrolling, which are however not directly usable to exploit the particularities of the SystemC constructs provided as a C++ library. Using these tools for SystemC necessitates the inclusion of the non-deterministic scheduler specification in the tools. Moreover, such tools do not usually take parallelism into account. For instance, CBMC [4-5] can apply bounded model checking techniques on pure C models, but cannot deal with parallelism or infinite loops. SLAM [6] uses clever abstractions and refinement techniques, but only focuses on sequential programs.

The closest related work is found in Java model checking, wherein a scheduler specification was taken into account. The Java PathFinder model checker, for example, successfully found bugs and proved properties on real programs. The first version [7] used an approach similar to ours: translating Java into the intermediate representation, Promela, and using the model checker SPIN to prove the properties. The second Version [8] checked the byte-code directly using a dedicated Java Virtual Machine (JVM) with backtracking capabilities and many other model checking techniques. However, the techniques dedicated to Java are *not* directly applicable either to SystemC and its scheduler or to the modeling of synchronous and asynchronous mechanisms.

9.1.2   Approach and Contribution

We advocate an approach to exploit all the particularities of a TLM design written in general SystemC. The method implemented in our new dedicated tool, LUSSY, is comprehensively described in the remainder of this chapter.

LUSSY is a tool based on the SystemC front-end Pinapa [9], which can extract architecture and synchronization information from a TLM design written in SystemC with very few abstractions. LUSSY builds its own intermediate representation, HPIOM (Heterogeneous Parallel Input/Output Machines), comprising communicating parallel machines that represent deterministic and non-deterministic components, synchronous and asynchronous communication protocols, Boolean and numerical data. For the time being, LUSSY connects this intermediate representation to the symbolic model checker LESAR [10] and the abstract-interpretation tool NBAC [11]. Both tools provide conservative automatic verification results for safety properties; they may perform their own abstractions on the HPIOM whenever necessary. The present state of the LUSSY implementation can already accept a large subset of SystemC. Currently, LUSSY is being applied to the case studies provided by STMicroelectronics.

Translating SystemC into HPIOM is a way of giving a formal semantics to SystemC. The faithfulness of the translation relies on the executability of HPIOM. The HPIOM obtained may be tested against the official SystemC execution engine. LUSSY is an open tool, allowing other tools (SAT solvers, model-checkers, etc) to be experimented on HPIOM obtained from SystemC.

The contributions of this section are threefold:

1. an executable formal semantics for TLM models written in full SystemC, with an operational translation tool;
2. a way of expressing safety properties directly in SystemC;
3. a working connection to verification tools.

9.2 Verification Approach

9.2.1 Verifying SystemC Programs

Our approach to verifying SystemC programs is underpinned by two important considerations listed hereafter.

- *Static and Dynamic Aspects*
The architecture is built by executing the *elaboration phase* of the SystemC program, which performs dynamic object allocations for all components and connections. The set of components is then recognized for all the execution time. The architecture being *static* is a crucial point: the set of processes is recognized once and for all, and the topology of the connections stays unchanged.

- *Synchronization Code vs Complex Algorithms*
A typical TLM design distinguishes clearly between the potential complex algorithms of certain components and the code dedicated for synchronization. For instance, a processor could be included in a TLM design with SystemC codes describing the interpreter of its machine language while the code intended to run on the processor is provided separately.

Our discussion will focus on the *safety functional synchronization properties* of TLM models (*safety* as opposed to *liveness* [12] and *functional* as opposed to *performance*). Such focus implies that a processor employed in a given design has to be treated in a very abstract way. The program run by the processor could be checked by other techniques such as software model checking or theorem proving. Then, the processor component may be replaced by very simple SystemC codes, describing its connections to other components as well as abstracting all its behavior. The properties that can be verified on such an abstracted TLM design are restricted from depending on the details of the algorithms run by the processor. Indeed, such restriction is a good design practice.

9.2.2   Expressing Properties to be Verified in TLM Designs

Generic properties do not require using a specification language. In LUSSY, the followings can be expressed and checked:

1. Verify that a global deadlock never occurs. We consider that a global deadlock occurs when the SystemC scheduler enters the "time elapse" phase while no process is waiting for time.
2. Verify that a process never finishes. This should always be the case except for test benches.
3. Verify that a synchronous signal is never written twice during the same delta cycle. This is a dangerous situation since the final value of the signal will depend on the order of execution.
4. Check concurrent accesses on TLM ports. Since concurrent accesses to TAC ports lead to undefined behavior, they must therefore be avoided.

In order to specify and prove user-defined properties of SystemC designs, specification formalism is required. The idea of LUSSY is that users should not have to learn a temporal logic language. The property should be written in the same language as the implementation.

The mutual exclusiveness of certain code portions may be verified. This is slightly intrusive in the source code because the beginning and end parts of the critical sections have to be specified.

Finally, the most general safety properties are expressed by assertions in the source code through ASSERT(condition). Technically, the macro is defined by `#define ASSERT(X) if(!(X)){is_this_reachable();}`. This reduces the assertion verification problem to the one of code reachability.

9.2.3   Didactic Example of TLM Design

To illustrate the transformation from SystemC to HPIOM, a simple example is introduced as depicted in Figure 5-24. For clarity, the example shows only the body of the processes and the methods called to process transactions in the slave modules. The program contains *assertions* for the properties to be verified.

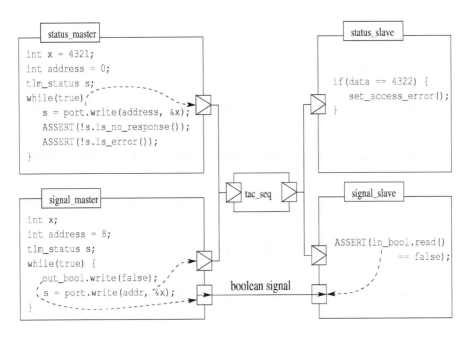

Figure 5-24. An Example of the Transactional System

The target port of the module `status_slave` (respective `signal_slave`) is mapped at the address 0 (respective 8). It receives the transactions initiated by `status_master` (respective `signal_master`). The call to the method `write` on an initiator port searches for a slave module mapped at the corresponding address, and calls the `WriteAccess` method on it. If no module is mapped at the address written on, then nothing happens and the status returned verifies `status.is_no_response()`. If a module is mapped at this address, the status returned verifies `status.is_ok()` unless the method `set_access_error()` has been called during the `WriteAccess` call.

In this example, the behavior highly depends on the value of variables representing addresses. The action triggered by the `write` will be totally different depending on the address on which we write. The behavior may also depend on the data because of the `if` statement, but there is no complex algorithm.

9.3   Semantics of SystemC in HPIOM

9.3.1   Principles

LUSSY uses an intermediate formalism called HPIOM. This formalism is expressive enough to allow an easy translation from SystemC, and simple enough as a formal semantics to allow easy conversion to other formalisms. HPIOM describes communicating synchronous automata. Each automaton has both explicit states and internal variables. At every instant, each automaton must perform a transition. A transition can be guarded by a condition. It can also trigger actions, i.e. assignments to change the value of a variable, or emission of signals that will be received in the next delta cycle.

By default, the translation into HPIOM does not perform more abstractions than those implied by the expressivity of HPIOM compared to that of SystemC (see section 9.3.2). Since most of the interesting properties are not decided on HPIOM, further abstractions will have to be made. However, this part is left for specific verification tools connected to HPIOM.

Since SystemC has no formal semantics, a formal proof of the equivalence between a SystemC source file and the corresponding HPIOM representation built by LUSSY is of course impossible.

The main idea is as follows:

1. each process in SystemC will be associated with one automaton in HPIOM, built from the information given by the C++ front-end;
2. the complete HPIOM description of a SystemC design will be made of all these "process" automata, plus specific automata for SystemC and TLM constructs.

We never parse the SystemC library source code itself but describe HPIOM patterns based on the SystemC library specifications. There is an automaton pattern for the scheduler; on top of it, there is an automaton for each signal, event, channel, etc.

9.3.2   Expressivity of HPIOM and Abstractions

HPIOM may be used to encode any statically bounded-memory program. In SystemC, static bounds are guaranteed provided that:

1. the program does not perform dynamic memory allocation;
2. there are no recursive function calls.

The semantics of translating SystemC into HPIOM abstracts memory allocation primitives and recursive function calls into new input messages with unknown values. The same is done for those SystemC constructs that are not yet implemented, in order to get a working connection to verification tools before full SystemC is taken into account by the front-end.

Another abstraction (which is optional) is related to the way addresses are dealt with. In our TLM models, addresses are simply `int` values. If nothing special is done in the translation, addresses become ordinary variables in HPIOM, and any property related to addresses has to be transmitted to a verification tool able to deal with `ints`. However, in the SystemC source code, it is possible to distinguish addresses from other `ints`. For addresses, we propose an encoding based upon the existence of *address maps*. Indeed, in SystemC, the relevant values of the address variables are given by the address maps describing the connection between components. Such a map is a partition of **N** into a finite number of *ranges*. A Boolean variable is associated with each of them. As soon as addresses are manipulated with this abstraction, we may lose information. This is conservative for safety properties. It simplifies the proofs a lot, and has proved to be sufficient on the examples we tried.

The last abstraction (which is also optional) is related to asynchrony. SystemC is intended to model and simulate *asynchronous* components. Although it provides a construct `wait(t)` where `t` is an amount of time, guidelines specify that this quantitative time `t` should not be used to enforce synchronization. In other words, the designer should not assume that two processes that perform the same `wait(t)` would synchronize when `t` has elapsed. The "time-elapse" phase of the scheduler algorithm wakes the processes up in the order specified by the `wait` parameters. In our translation, they are awakened non-deterministically (encoding non-determinism with oracles). It means that the HPIOM model exhibits more behaviors than what the SystemC interpreter does. This conservative abstraction enforces the guidelines: if a safety property can be proven on the HPIOM model, then it is true that the `wait` statements have not been used to enforce the synchronization.

9.3.3   Semantics of Translating Process Code into HPIOM

Compiling imperative codes into automata is a widely known problem and there is no semantic difficulty here. However, the abstract syntax tree for a C++ design contains many particular cases. Among these cases, a lot of them must be taken into account if we want to apply our tool to the real practical SystemC designs. Hence, such translation represents a significant part of the implementation work. The control flow of a `while` loop is given in Figure 5-25 as an example.

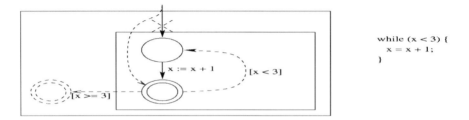

Figure 5-25. Control Flow of a While Loop

9.3.4   Semantics of Synchronization Primitives and Scheduler

Expressing the semantics of the scheduler by some synchronization between the HPIOM automata could be accomplished in several manners. The global communication scheme is illustrated in Figure 5-26 and will be detailed immediately. The semantics of the SystemC scheduling policy is modeled by an automaton for the scheduler. In addition, two automata are modeled for each process: one represents its control structure (as explained earlier) while another represents its state in the scheduler (as depicted in Figure 5-27). The process could be in one of the following states:

1. running;
2. ready to run (i.e. eligible);
3. sleeping (blocked by a wait statement for a *SC_THREAD* or execution over for an *SC_METHOD*).

The synchronization between the two automata is such that the first automaton (representing the control structure) may change its state only if the second one is in the state of "running".

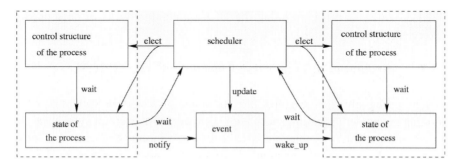

Figure 5-26. Global View of the Communication between Automata in HPIOM

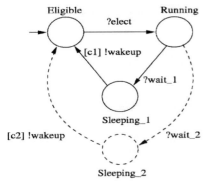

Synchronizations:
– elect: received from the scheduler when this process is elected
– wakeup: sent to the control structure
– wait_2: received from the control structure when a wait statement is reached
– c1 and c2 correspond to the conditions conditions the process is waiting for in the corresponding "sleeping" state

Figure 5-27. State of a SystemC Process

The SystemC scheduler itself is represented by an additional automaton as illustrated in Figure 5-28. It starts in the state of "selecting_process". At that particular starting moment, all of the processes are eligible. The SystemC official definition lets the choice among the eligible processes unspecified. In our modeling approach, the scheduler chooses a process non-deterministically; meaning that when we prove a property of a SystemC design including this non-deterministic scheduler, we prove it for any possible implementation.

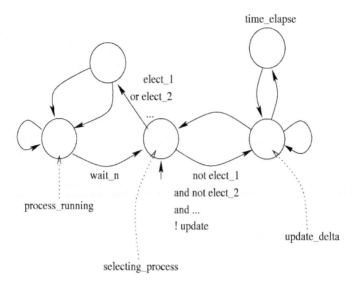

time_elapse

elect_1
or elect_2

...

wait_n

not elect_1
and not elect_2
and ...
! update

process_running

update_delta

selecting_process

Synchronizations:

– elect_n: sent to the corresponding process state automaton and control structure

– wait_n: received from the corresponding process state automaton

– update: sent to all processes that may have an action to execute during the update phase

Figure 5-28. The Pattern of SystemC Scheduler

The low-level synchronization primitive in SystemC is called `sc_event`. The operations available for a given `sc_event` include:

- `notify()`
  The event is triggered immediately.

- `notify(SC_ZERO_TIME)`
  The event is triggered at the end of the delta-cycle.

- `notify(time)`
  The event is scheduled to be triggered for some date in the future.

We also build a HPIOM automaton for each `sc_event` according to the pattern depicted in Figure 5-29. It has an initial state and a state for each kind of delayed notification. The immediate notification is modeled by a single transition. In any case, the transition going back to the initial state is the transition triggering the event. It emits a message that will move processes waiting for that event from "sleeping" state to "eligible" state.

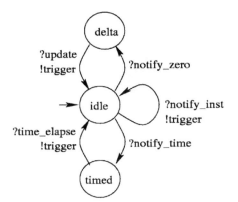

Synchronizations:
– notify_... messages are received from the control structure when a notify statement is encountered,
– trigger goes to the process state automaton, and will make the condition to the "eligible" state true,
– update and time_elapse both come from the scheduler.

Figure 5-29. Pattern of sc_event

9.3.5 Direct Semantics of TLM Constructs

Although TLM constructs, as mentioned earlier, are library components whose codes could be translated using the translation schemes above, we advocate a translation wherein these constructs are given a direct semantics in HPIOM. It allows exploiting the information they give on the structure of the design. In this section, we sketch the HPIOM encoding of TAC Channel.

- *Wait for the channel to be available.*

For each master port, we first create a "waiting" automaton synchronized with the master and the TAC. It simulates the master process waiting for the TAC to be available as in Figure 5-30. If the TAC is unavailable when a transaction is initiated by a master, the master should let other processes run. It will become eligible again when the TAC selects its transaction.

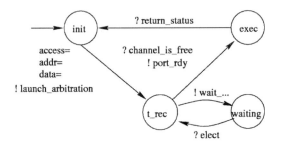

Synchronizations:
– channel_is_free is received from the channel when it is available.
– wait and elect are communication with the scheduler to wait until channel_is_free is present.
– return_status is received when the transaction processing is over.

Figure 5-30. Wait for Channel Availability

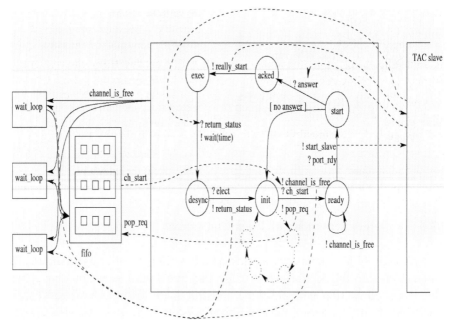

Figure 5-31. Pattern of tac_seq

• *Select transaction and resolve the address.*

The TAC itself is modeled by the automaton as depicted in Figure 5-31. It loops in the initial state until it receives a transaction. When transactions are ready to be executed, values identifying them are entered in a FIFO (finite FIFOs are encoded into HPIOM). The automaton of the channel processes these values one by one and goes to the state of "ready". A message is sent to all the automata modeling slaves, and those whose address maps match the answer. If the channel gets no answer, then it returns immediately with a status is_no_response. In the previous example illustrated by Figure 5-24, the first process elected will send the first transaction that will be processed immediately, and the next one will queue until the transaction is processed.

• *Execute the corresponding method in the slave module.*

When the transaction is selected and the slave is identified, the body of the corresponding method in the slave module is executed and the status is returned (i.e. the state of "exec" as shown in Figure 5-31). Note that the scheme is a somewhat simplified here since the channel has to communicate with several instances of slave modules.

• *Simulate a **wait** to allow other processes to execute.*

If a slave module responds, the automaton of Figure 5-31 will simulate a wait statement (i.e. the state of "desync") for a given time duration.

9.4 The Tool LuSSY

The tool LuSSY has an internal structure similar to the one of a compiler as in Figure 5-32. The front-end extracts information from the system; a second pass compiles it into an intermediate representation called HPIOM, taking the semantics of SystemC into account; and a code generator gives a textual representation for it, which can be used as input by other tools.

The HPIOM representation can easily be converted into several formats usable by external tools. To date, we have a LUSTRE back-end that permits the utilization of LESAR and NBAC. We also have two visualization back-ends, one for viewing the connections between automata and another for viewing the automata themselves, using the *dot* format of the graphviz[10] package.

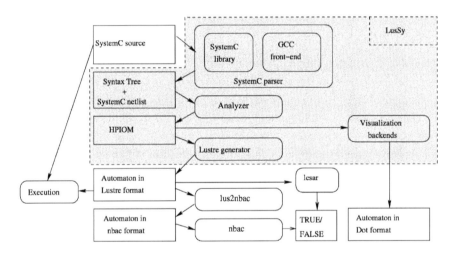

Figure 5-32. LusSy Tool Chain

9.5 Applying LuSSY to the Example

Let us get back to the example of section 9.2.3. In the module `signal_master`, we write a value on the channel at the address 8 after writing the value `false` on a signal. The module `signal_slave`, mapped at this address, will receive the transaction, and check that the value of the

---

[10] Refer to the web site of graphviz for further information: http://www.graphviz.org.

signal is `false`. This may seem trivially true, but it is *not*. The semantics of `sc_signal` says that the value is actually taken into account only in the next delta-cycle. As there is no `wait` statement between the `write` and the `read` statements, the value read is the previous value. During the first iteration of the loop, the value read is the initial value of the signal. In practice, with the current implementation of the SystemC library, the value is initialized to `false`. It is however clear from the SystemC specifications that the initial value is unspecified. The bug is consequently masked during the simulation, and NBAC cannot prove the property. The diagnosis of the proof failure gives the condition on the initial value of the signal (`true`) that causes the bug. If we explicitly initialize the signal to `false`, then the property becomes provable; if we explicitly initialize it to `true`, then the property is false and the assertion is actually violated during the execution.

After identifying the bug, we can fix it, for example, by adding a `wait` statement:

```
while (true) {
    out_bool.write(false);
    wait(SC_ZERO_TIME);
    status = master_port.write(address, x);}
```

Once the assertion is verified, NBAC is able to prove the correctness of the assertions. Now, look at the module `status_master`. It just writes on the channel, and tests the status returned. If a value not equal to 4322 is written at a mapped address, then the property is true and provable by NBAC (but not by LESAR since the property is data-dependent). The first assertion becomes false if the address written to is changed, or the address map is altered, making it write on an unmapped address. If the data 4322 is written, the slave will set the error flag; the second assertion becomes false, and the proof fails.

Indeed, this example has the complete verification flow. The proving tool is able to prove true properties with no manual intervention in less than a second. The model contains 28 automata in parallel, with a sum of 104 states, 196 transitions, 10 numerical variables, and 50 Boolean variables. This may seem quite a lot compared to the source size, but the global state-space is not as large because the system is made of many tightly synchronized small automata. Knowing that optimizations must address the number of variables before anything else, we use symbolic tools for which the key point is the number of variables but not of global states.

9.6   Conclusion on Formal Verification

This section has presented our approaches and tools for the analysis of SystemC transactional models. Starting from the source code of a SystemC design, it is parsed using GCC's C++ front-end and the SystemC library itself, then transformed into a set of automata, and finally dumped into the Lustre language. The implementation is operational and the faithfulness of the translation was validated on basic examples by comparing the executions of the generated Lustre to the executions of the "official" SystemC implementation.

The connections to two different model checkers that did not perform the same amount of abstractions were also conducted. The main idea of the approach is to extract as much information as possible from the SystemC design, and let the verification tools perform the abstractions required.

We are currently applying the whole approach to a significant case study provided by STMicroelectronics, in order to identify the optimizations of the encoding. On HPIOM, we also experiment some traditional compiler techniques such as variable analysis.

LUSSY, being an open tool, can easily move to another model checker by rewriting the translation from our automata to its input format. We are starting experimentations on SMV and SAT solvers. One of the easiest attempt is trying out the tool by Prover Technologies. Lustre is indeed the basis of the SCADE environment provided by Esterel Technologies, which is equipped with a plug-in by Prover Technologies AB providing SAT solving and Presburger arithmetic. Our translation of HPIOM into Lustre can be directly used in SCADE.

As LUSSY provides a formal semantics of SystemC, it can be the basis of a toolbox for the development of SoCs at the transactional level. In other words, LUSSY can provide tools for all the questions related to TLM design, ranging from verification and test at the TLM level to the comparison of TLM and RTL levels, and analysis of non-functional properties. For instance, the formal semantics can be employed as a support for the automatic generation of test sequences intended to run on both the TLM and RTL models of a given design. This reference semantics is also the necessary starting point for comparing executions at different levels of abstraction.

Since HPIOM preserves the potentially complex algorithms of SystemC codes, powerful software verification techniques could be used (e.g. invariant extraction, predicate abstraction, etc.). Besides, we could consider extending HPIOM to manage dynamic data-structures, but this would require efficient support in the proving tools. A more promising approach is the systematic use of *contracts* for some of the components. As mentioned earlier, a processor is an interpreter of the binary code. It has to deal with a complex data, i.e. the C codes to be executed! C codes should be abstracted, i.e. the values exchanged are replaced by unknown values encoded by inputs. However, such abstraction requires some assumptions of the processor behavior concerning the way it synchronizes with other components. This can be described by a contract.

REFERENCES

[1] K.L. McMillan, Symbolic Model Checking, Boston: 1993.

[2] R. Drechsler and D. Grosse, "Formal Verification of LTL Formulas for SystemC Designs," in Proc. of ISCAS, 2003.

[3] T. Sakunkonchak and M. Fujita, "Verification of Synchronization in SpecC Description with the Use of Difference Decision Diagrams," in Proc. of the Forum on Specification & Design Languages (FDL'02), 2002.

[4] E. Clarke and D. Kroening, "Hardware Verification using ANSI-C Programs as a Reference," in Proc. of ASP-DAC 2003, 2003, pp. 308-311.

[5] E. Clarke, D. Kroening, and F. Lerda, "A Tool for Checking ANSI-C Programs," in Proc. of Tools and Algorithms for the Construction and Analysis of Systems Conference (TACAS'04), 2004, pp. 168-176.

[6] T. Ball and S.K. Rajamani, "Boolean Programs: A Model and Process for Software Analysis," Microsoft Research, Feb 2000.

[7] K. Havelund, "Java PathFinder: A Translator from Java to Promela," in Proc. of SPIN, 1999, pp. 152.

[8] W. Visser, K. Havelund, G. Brat, and S. Park, "Java PathFinder - Second Generation of a Java Model Checker," in Proc. of Post-CAV Workshop on Advances in Verification, 2000.

[9] M. Moy, "Pinapa: A SystemC Front-end," Online Open Source Software and Manual, Available at: http://greensocs.sourceforge.net/pinapa/

[10] N. Halbwachs, F. Lagnier, and C. Ratel, "Programming and Verifying Critical Systems by Means of the Synchronous Data-Flow Programming Language Lustre," IEEE Transactions on Software Engineering, 1992.

[11] B. Jeannet, "Dynamic Partitioning In Linear Relation Analysis. Application to the Verification of Reactive Systems," Formal Methods in System Design, vol. 23, no. 1, pp. 5-37, 2003.

[12] L. Lamport, "Proving the Correctness of Multiprocess Programs," IEEE Transactions on Software Engineering, vol. SE-3, no. 2, pp. 125-143, 1977.

Chapter 6

ARCHITECTURE ANALYSIS AND SYSTEM DEBUGGING
A Transactional Debugging Environment

Antoine Perrin and Gregory Poivre
STMicroelectronics France

Abstract: Given the complexity of SoC development in the nanotechnology, it has become critical to fully validate the system performance at the early stage of the SoC design flow. This chapter describes the tools and methods for evaluating the overall SoC interconnect performance, for which the commercial solutions are not yet available. The proposed methodology is based on SystemC simulation using a generic IP Traffic Generator (IPTG) and a powerful monitoring mechanism called SysProbe, which are applicable all through the SoC analysis flow ranging from the transactional to register transfer level (RTL) simulations. Such Traffic Generators model the system IPs and the system traffic dependency with a refinement flow, while real slaves or targets are used to generate the correct latency. The SoC architecture is modeled either at the transactional or RTL level according to the requirements of development costs, simulation speed and precision. SysProbe provides the results of the architectural analysis to SoC architects.

Key words: transaction; architecture analysis; architecture platform; transactional debugging; monitoring; transactional viewer; IP traffic generator; SysProbe; traffic characterization; configuration file; initiator; target; interconnect; communication model; memory structure model; cycle accurate model.

1.   DEFINING SYSTEM-ON-CHIP ARCHITECTURE

1.1   Architecture Definition

Defining a SoC architecture and micro-architecture that will sustain the real-time constraints of the targeted application is a great challenge. It is yet

F. Ghenassia (ed.), Transaction Level Modeling with SystemC, 207-240.

again another challenge to verify whether such an architecture or micro-architecture fulfils the target real-time constraints.

Assume that every IP of a SoC is sustaining its real-time constraints, the architecture/micro-architecture definition and verification with respect to the SoC performance must then focus on the following critical components:

- communication structures[1];
- shared memory controllers.

To help define these communication and memory structures, an environment comprises the appropriate tools, models, and the associated method must be made available. This environment addresses not only the SoC architects working on the communication and memory structures, but also the verification engineers verifying the compliance of the SoC implementation with the application constraints.

Two main input categories are distinguished for this environment:

1. *IP Traffic Characterisation.*
   Every SoC IP that influences the architecture definition must be modeled in terms of the traffic it generates.
2. *Application Real-time Constraints.*
   A given SoC targets a specific application or application domain. The real-time constraints associated with this application must be made accessible so that the estimated or measured SoC performance can be compared to the performance results analyzed using the application constraints.

The greatest challenge to implementing the methodology based on the above environment is getting SoC architects, who currently use spreadsheets to define the SoC architecture, to adopt this new approach. The appropriate solution must therefore propose a very simple iteration cycle loop without obligating the users to learn a new debugging language. The number of components should also be reduced to the minimum by eliminating those components that have no direct impact on the performance analysis and system debugging.

The simplification of the SoC platform assembly can be attained by adopting the SPIRIT automation strategy and the SPIRIT compliant tools (see Chapter 7). The SPIRIT automation flow is a compulsory pathway to implementing the new methodology described in this chapter, i.e. transactional architecture analysis and system debugging. This automation flow combines SoC components of various abstraction levels with high

---

[1]   Including FIFOs (first-in-first-out) used to access to the communication backbone.

efficiency. The next section gives the list of the components constructing this environment.

1.2   Components of Architectural Platform

The transactional debugging and architecture analysis environment is composed of the following components:

- *Analysis Tool (AT)*
  AT monitors a simulated system in order to provide the results that are directly related to the target application constraints.
- *Intellectual Property Traffic Generator (IPTG)*
  The IPTG reads a configuration file describing an IP in terms of its traffic in order to re-generate the corresponding traffic on the communication backbone. Advantages of using generic traffic generators include avoiding time delay due to unavailable models, easier maintenance than C models, and direct accesses to the traffic scenario for validation.
- *Communication Model (COM)*
  The COM models the communication backbone of a SoC platform. It serves during the analysis phase to help define the communication micro-architecture features such as topology, arbitration, and FIFO size.
- *Instruction Set Simulator (ISS)*
  The ISS is used for three cases. First, communication structures and memory controllers are often programmed by a processor. During the architecture analysis, the ISS is used to perform this programming task. Second, the ISS is frequently used for the analysis of interrupts. Third, the traffic generated by the processor must be taken into account as well. The ISS can handle this task adequately. While the usage of ISS is obligatory for the second purpose, the other two purposes can be served just as well by using a generic traffic generator.
- *Bus Functional Model (BFM) or Transactor*
  The BFM or transactor establishes and assures the correct integration and communication between components of different abstraction levels. This component allows a progressively refined model to be easily integrated into a SoC platform throughout the design cycle, for instance, starting from TLM IP, to BCA IP, and finally RTL IP.
- *Memory Structure Model (MEM)*
  The MEM models the memory controller and the memory module with enough details to accurately represent the access latency.

The components introduced above define the overall architecture of a SoC platform as depicted in Figure 6-1. These components can be applied in a modular manner to adapt for the specific context of a SoC design team.

A generic approach to the SoC performance analysis and verification is described hereafter as the starting point for defining specific approaches.

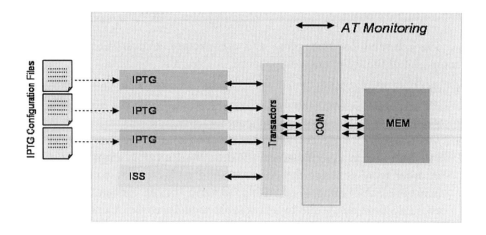

Figure 6-1. The Architecture of a SoC Platform

Three phases are undertaken as generic approaches, starting from the early definition of the main SoC architectural components down to the verification of the real chip performance.

1. *Early Micro-Architecture Definition.*
   This definition is based upon IPTG, COM and MEM models. The RTL model of the memory controller and the behavioral hardware description level (HDL) memory model could probably be used if the abstract models are not available. The analysis environment is provided by the AT. This phase aims at defining the major micro-architectural SoC features such as topology, FIFO size, arbitration, and IP clustering.

2. *RTL Performance Verification.*
   This verification is based on the IPTG and RTL implementations of the components under study. The AT computes the performance figures and compares them to the same features estimated during the early micro-architecture definition. The IPTG configuration files applied in the first phase are reused here to generate the identical traffic. Indeed, this phase verifies if the communication and memory models are in compliance with the equivalent RTL implementation.

3. *On-Chip Performance Verification.*
   The third phase is based on the real chip. During the chip verification, an on-chip performance monitor extracts traces from the chip activity.

These traces are given as input to the AT that will subsequently compute the performance figures and compare them with the measured performance results on the RTL implementation. Some discrepancies will be noticed because the traffic is generated by the real IPs in this phase while it is generated by IPTG in the RTL model. This comparison is very useful to understand how accurate the IPTG configuration files are with respect to the real IP traffic. The third phase verifies the accuracy of the IPTG versus the real IPs in the real context.

2.   TRANSACTIONAL DEBUGGING

2.1   The Need for Transactional Debugging

The current SoC generations are based on the multiple initiators/masters and the multiple targets/slaves. A powerful routing system is required to interconnect all of these IP blocks, for instance, OCP [1], STBus [2], and AMBA3.0 [3]. The efficiency of a routing system in conducting the performance analysis depends strongly on the functionality and the programming of the system.

Today, the routing system has two drawbacks. First, the complexity of the routing system continues to grow exponentially. Such growth makes it impossible to carry out the conventional manual traffic analysis and architecture study on paper. Although this manual analysis continues to be helpful in defining the basic system architecture, a simulation tool must be used to perform a complete traffic analysis and architecture study. Second, the routing system may result in a system with mixed frequencies and a huge number of IP instantiations of mixed protocols during the simulation of the system integration.

If a problem occurs during the SoC integration, engineers will need to check all components of the SoC platform simultaneously. This could be a tedious and lengthy job. Consider that a routing connection includes 20 signals in average. If the integration problem occurs, engineers might have to check up to thousands of signals! Each of these signals represents one line in a traditional waveform viewer. Bear in mind that a real signification of these signals can only be interpreted by combining several signals.

All the problems described above have raised the need for an efficient solution. This chapter describes our methodological approach, *transactional debugging*. The principle of the transactional debugging lies in the transformation of such signal combinations into a unified transaction, with the intention to collect all the necessary information in the same location.

By adopting the transactional debugging methodology, the debugging effort is significantly reduced. In the example above, a direct advantage is the reduction of 20-line simultaneous cross-check in a waveform viewer per signal trouble-shooting.

Moreover, the transactional debugging helps to avoid the typical lengthy and tedious study of the different bus protocols for understanding the bus communication in a system. Not only are time and efforts saved, but the analysis results of the transactional debugging are much more user-understandable and user-interpretable than those of signal analysis.

Another interesting advantage of the transactional debugging is that all types of protocols and abstraction levels could have the same representations and attributes. In addition, there is at least a common set of parameters made available for all kinds of point-to-point connections. The rest of the parameters and transactional structures are defined by the communication structure.

2.2   Definition of Transactional Debugging

Before getting into further details of the transactional debugging, certain conceptual definitions are briefly described in this section.

A transaction is defined as a unified element representing a set of data being exchanged. It includes a list of parameters with each characterized by its name and value. These parameters can be called later as attributes of the transaction. The transfer of a transaction is denoted by a starting and ending date.

A transaction stream is a set of transactions occurring under a particular context. For instance, transactions between two routing interconnections are grouped as a specific transaction stream. According to the interconnection properties, transactions can be overlapped. An overlap of transactions occurs when a transaction starts its transfer on a stream before other transactions previously stored on the same stream end their transfers. To indicate the hierarchy between different transactions, logical relations can be defined to represent their inter-relations; for instance, predecessor-successor or parent-child relations.

The transactional information is fully compatible with the corresponding signal information. Both of them can thus exist in the same environment.

2.3   Transactional Debugging Environment

The transactional debugging is essential for the current SoC generations. It raises the observation level from signals to transactions, and thus reduces the complexity of the interconnection or communication representation.

To apply the transactional debugging environment in the SoC analysis, some fundamental building blocks are required. First, monitors for all interconnections of different abstraction levels must be made available. Second, an environment supporting the transactional debugging needs to be set up. Therefore, the AT must be equipped with a set of monitors and an analysis environment for transactional debugging. Such AT environment is called *SysProbe*, standing for System Probe.

The AT monitor is a Finite State Machine (FSM) that recognizes the protocol of the communication structure in a SoC platform for extracting information such as addresses and data transferred. The AT analysis environment should support the recording, visualization, and analysis of transactions. It should nevertheless be able to mange traditional signals too.

2.4   Monitoring Principles

In the transactional debugging, monitors are made available on a given SoC platform for:
- different abstraction levels of the same communication structure;
- different communication structures.

Such monitors are built in different manner according to the associated abstraction levels. Natively available in TLM, the monitor is only surveying the actual TLM interface or communication function. Quite opposed to the idea of cycle accurate monitoring as depicted in Figure 6-2, the transactional monitor is composed of the following components:

1. Data Acquisition Components
   Either group of modules listed below is in charge of data collecting:
   a) Simulator Link Layer. Its role is to assure the connection between the monitor and the simulator for a dynamic transaction recording. The key advantage of such layer is to obtain transactions during the simulation runtime session. The only disadvantage is that the monitor must be manually instantiated before launching the simulation. Through the SPIRIT design automation, this part will be fully automated and transparent for end users.
   b) Value Change Dump (VCD) File Parse. This module is another option for data acquisition that is used in the post-processing

mode. This method is not interactive because the results can only be studied at the end of the simulation.

2. Finite State Machine

This module is responsible for extracting the signal information collected by data acquisition components, and processing them into the transactional information according to the associated bus protocol before sending them to output modules.

3. Output Modules

There are several output modules for handling the simulation output:

a) Transaction Dumper. This module obtains the transactional information from the finite state machine, prepares them into the final database formats, and dumps them into the database.

b) Protocol Checker. This module has two missions. Its first mission is to assure the transaction integrity by detecting protocol violations that may affect the attribute integrity of the information. To perform its first mission correctly, the protocol checker must be able to verify a minimum set of the protocol rules. Thus, the module is actually performing its secondary mission to verify partially the protocol compliance.

c) Performance Analyzer. The analyzer records the native information of the transaction such as latency, frequency, occupancy, etc.

d) Transaction Linker. Based on a specific algorithm, the linker is in charge of detecting the relationship between all transactions to deduce a "system-level link" between all transactions.

e) Traffic Generator. This module creates a configuration file that could be reused by IPTG.

The following are the main performance figures logged for performance evaluation of a given SoC platform:

1. latency statistics;
2. pipeline statistics;
3. opcode distribution;
4. occupancy;
5. throughput;
6. bandwidth;
7. bandwidth occupation.

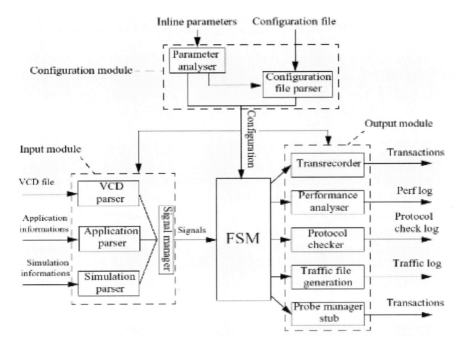

Figure 6-2. Cycle-Accurate Monitoring Structure

As illustrated in Figure 6-2, the FSM is implemented in C++. Note that the recording mechanism is implemented using SystemC Verification[2] (SCV) library. This standard improves the inter-operability between ATs by providing APIs for the transaction-based recording. Monitors used in the transaction recording allow targeting all database formats (whose recorders implement the SCV transaction dumping API with the same code through the unified API provided by SCV). To manage a new SCV compliant database format, the only action to perform is to link the new recording library with the existing probe. Thus, designers can use their own analysis environment such as text based, Cadence Incisive [4], or Novas Verdi [5].

2.5 Analysis Environment

The transactional debugging analysis environment consists of two parts:
• waveform viewer;
• user plug-in with query and add-on debugging features.
Further details on both parts are provided in the following sub-section.

---

[2] SCV is the extension of SystemC for verification.

2.5.1   Transactional Viewer

The real-time debugging is fully linked with the high capabilities from the AT to display transactions along with traditional signals. For this reason, all the traditional operations applicable to the signal display should also be applicable to the transactional display. Typical examples of such include comparison, search, splitting transaction attributes (analogous to splitting signals in a bus explosion), and expanding all transaction events occurring at a particular time instant.

On top of these basic display functionalities, a transactional viewer must take into account various aspects of transactional structure. It means that the viewer should provide the capabilities of displaying transaction overlaps and transaction attributes (with flexible control over which attributes to display). Cadence Incisive and Novas Verdi are two powerful tools that support the transactional display with high efficiency. Figure 6-3 shows the example of SimVision transactional display from Cadence Incisive.

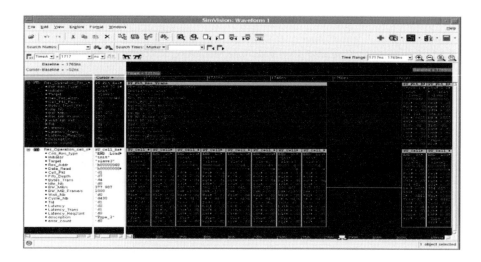

Figure 6-3. SimVision Transactional Display

2.5.2   Viewer Statistics Plug-in

The transactional monitor is delivered with a set of predefined queries, SysProbe Analysis Generator (SPAG). These queries are automatically generated for a given design based on a configuration file that depends on the communication structure, COM. All of the COMs supported by the AT integrate a set of SPAG facilities. The results collected by these queries serve as the basic traffic statistics for that design. Three groups of SPAG queries are available as shown in Table 6-1.

Table 6-1. SPAG Query

Query Type	Query Function
Generic Query	Queries independent of COM.
COM Generic Query	COM-dependent queries for full COM analysis.
Project Query	Project-dependent queries.

Based on the viewer statistics, various performance evaluations can be examined. As an example, typical analyses obtained through the Cadence Incisive environment are (see Figure 6-4):

1. COM bandwidth;
2. COM opcode distribution;
3. COM memory map access;
4. COM memory map bandwidth;
5. COM latency statistics;
6. Initiator COM map access;
7. Link between query database table and wave viewers.

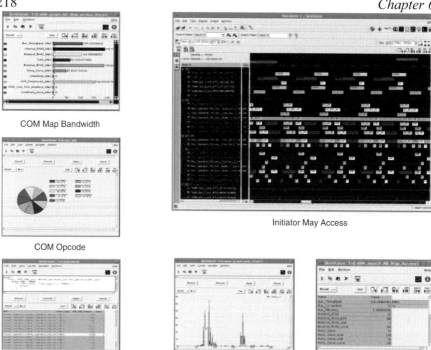

COM Map Bandwidth

COM Opcode

Initiator May Access

COM Access   COM Bandwidth   Latency Statistics

Figure 6-4. Cadence Incisive Statistics Plug-in Environment

2.5.3   User-defined Statistics

In addition to the SPAG query set, users are allowed to define their own set of queries. The user-defined query is based on the Cadence Incisive tool, Transaction Explorer (TxE) [6]. This tool provides users with an easy way to create specific queries by using the "browse button" where options are proposed at every step of a query creation.

2.5.4   Embedded Software Plug-in

SysProbe is delivered with a software-profiling plug-in called SysProbe Embedded Software (SPES). This plug-in associates the embedded software with the program counter (PC) signal of a given design to allow performing the hardware-software analysis. Results collected by SPES serve for creating a correspondence between the software execution and the hardware transactions recorded on the system.

Based on the disassembled code, SPES creates a correspondence between source codes and assembly codes during the debugging process to enable following the code execution by tracking the PC value. Although these principles seem similar to those used in common debuggers, SPES provides complementary results that give additional benefits to post-mortem analyses such as capabilities of moving at arbitrary execution time, going back during execution, and software profiling.

SPES has two key features:

1. *Software Execution Display*
   Note that SPES is not a debugger. As illustrated in Figure 6-5, SPES provides a post-processing tool that allows hardware designers to understand the software execution without adding an ISS.

2. *Early Software-Profiling*
   SPES is used for profiling early software execution, particularly in analyzing the functioning of the interrupt request (IRQ) for a given SoC platform.

Typical services offered by SPES include:

1. Correspondence between the time cursor and executed source codes.
2. Correspondence between C and assembly codes.
3. Display of function calls as signals.
4. Replacement of bus opcode signals by corresponding function names.
5. Duration and frequency of function calls.
6. Execution number of a specific code line.

Instead of providing just a simple probe, the capabilities listed above can be extended to provide a system view that relates the embedded software to the whole system. Thus, SPES makes it possible to track the execution of software commands in a system. Such ability allows analyzing the software performance according to the system architecture, and determining the arbitration influence on the speed of software execution.

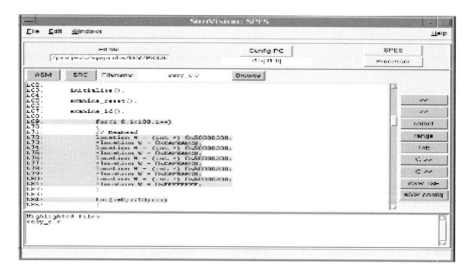

Figure 6-5. Embedded Software Plug-in of Cadence Incisive Environment

2.5.5 Transactional Link Plug-in

Transactions are characterized by particular relationships among them. The analysis environment of the transactional debugging must support features that describe the inter-transactional relationship.

The transactional viewer presented in section 2.5.1 supports such feature. The transactional linker in the transactional monitor works very well in a system where all the monitors can communicate together through the same bus protocol.

This feature, however, is not supported in certain cases where several heterogeneous systems coexist. A typical example is the simulation with mixed abstraction levels. A specific plug-in is therefore developed to handle this situation. As depicted in Figure 6-6, this transactional link plug-in creates the virtual link between transactions using a post-processing engine. This added feature is particularly useful to follow up the life cycle of transactions. By tracking the transactional life cycle, any resultant transaction during the simulation will be traced from its creation to its ending. A currently unavailable feature is a tracking from transactions to the resultant signals.

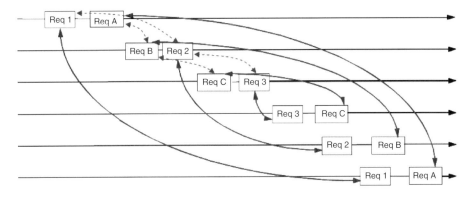

Figure 6-6. Example of Transactional Flow Link

2.6 Verification Role of Transactional Monitors

On top of its role as an analysis tool, the transactional monitor also serves as a verification tool for TLM IPs. The principle of such verification methodology is described hereafter.

To begin with, RTL signals of an IP under test are extracted and converted into transactions from an RTL test bench Based on this information, the transactional monitor will generate a set of IPTG configuration files.

Subsequently, a platform comprising an IPTG, COMs, and the TLM model of the IP under test is constructed for validation. According to the IPTG configuration files generated earlier by the transactional monitor, the IPTG will generate the same traffic as monitored at the RTL level. Through the comparison facility of the AT, the simulation results of the TLM IP are compared to those of the RTL IP for verification purposes.

2.7 Comparison of Abstraction Levels

Communication structures become similar in a certain sense as they approach the transactional level. At this point, including a subset of similar information in the communication structure will help to detect easily the discrepancy between different levels of abstraction.

A specific tool, *TransCompare*, is developed based on this concept. This tool computes the divergence percentage and lists all the discrepancy points of two traces. Such analyses can be purely functional or timed. Indeed, the engine of *TransCompare* ignores the timing information. The timing information is actually treated the same as any other transaction attributes.

By allowing the user to select on which attributes a computation is performed, *TransCompare* provides a direct access to both a pure functional and timed comparison tool. In addition to this key role, *TransCompare* is able to align the different naming conventions of transaction attributes from different database. This feature allows the transaction attributes or parameters to be correctly identified for comparison.

The main advantage of *TransCompare* is that it considers the transaction as a data flow by extracting timing information as parameter or attribute of the transaction. For this reason, this tool permits computing the functional convergence of different transactions even if their timing is completely irrelevant. An interesting added value of *TransCompare* is its transaction-filtering mechanism. Considered as data flows, transactions are easily filtered according to their attributes. Through this filtering mechanism, transactions traced from an IP can be compared to its reference even if it is in the integration phase. This method is also fully applicable to the emulation traces using the VCD input features of the monitoring tool.

3.   TRAFFIC GENERATOR

As introduced in section 1.2, the intellectual property traffic generator (IPTG) is a critical component in a given SoC architectural platform. The IPTG is a SystemC block that reads a traffic characterization file (i.e. IPTG configuration file) as input, and subsequently re-generates the corresponding traffic as output on the platform communication structure.

3.1   Principles

The IPTG is instantiated in a SoC platform following the same manner of instantiating any other components. The ultimate goal of having an IPTG instantiated for a given IP is to generate the traffic specific for that IP on a SoC platform.

A typical SoC platform incorporated with the IPTG could include the components at any of the abstraction levels listed below:
1. timed transactional level modeling (timed TLM);
2. bus cycle accurate (BCA);
3. register transfer level (RTL).

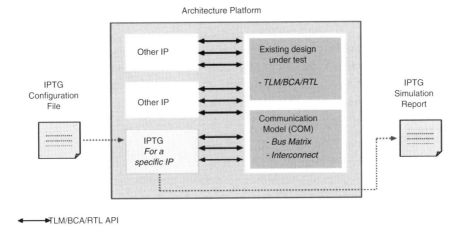

Figure 6-7. SoC Platform with IPTG Instantiation

The structure of a SoC platform with an IPTG instance is depicted in Figure 6-7. Note that the design under test shown in the figure represents an IP or a subsystem under test. Once instantiated in a SoC platform, the IPTG generates traffic on the ports of the communication model, COM. The COM ports are coded at one of the three different abstraction levels mentioned earlier: timed TLM, BCA or RTL.

As shown in Figure 6-7, the input to the IPTG is a configuration file that holds the following information:

- full statistical traffic;
- optional refinement;
- opcode sequence list;
- IP characterization parameters such as frequency and data size.

According to the information of the configuration file, the IPTG re-generates the IP traffic as the output. Another interesting feature of the IPTG is that a simulation report of the traffic generation could be produced by the IPTG for observation. In addition, a synchronization mechanism is implemented in the IPTG to model the dependency between system events.

3.2 Core Implementation

The building concept of the IPTG is based upon the standard of Open SystemC Initiative (OSCI). It is therefore fully compatible with the tools of the mainstream EDA providers. Furthermore, the IPTG is equipped with the randomization capability founded on SCV, which is an extension of SystemC for verification. Since both OSCI and SCV are open sources, the IPTG is a tangible solution totally free of charge.

3.3   Traffic Characterization

A given IP can be considered as a succession or a series of synchronized processes. The IPTG considers any single process or any group of these processes as *behavior*. An IP, therefore, is described by the IPTG as a series of behavior where each of them represents a particular type of IP traffic.

There are two approaches to define and model the characteristics of the IP traffic:

- *Traffic Modeling*. Define an IP by a set of behavior where each behavior represents specific bus traffic as seen from the external world. The IP is therefore viewed as a black box by users. The overall bus traffic of the IP could be considered as different specific traffic pieces that represent different IPTG behavior. The detailed information to configure these traffic characteristics is specified in an IPTG configuration file. In addition, there is a rather simple block to ensure a good consistency for all the behavior switching and overlapping.

- *IP Modeling*. Define an IP by a set of behavior where each behavior represents a specific internal IP traffic. This internal traffic is managed by a bus plug-in interface to subsequently create the bus traffic. The bus plug-in interface is represented by a FIFO with a threshold value and an opcode list. As illustrated in Figure 6-8, the bus traffic generated by the IP is split into two parts: (i) IP traffic that fills the FIFO, and (ii) FIFO traffic on the bus.

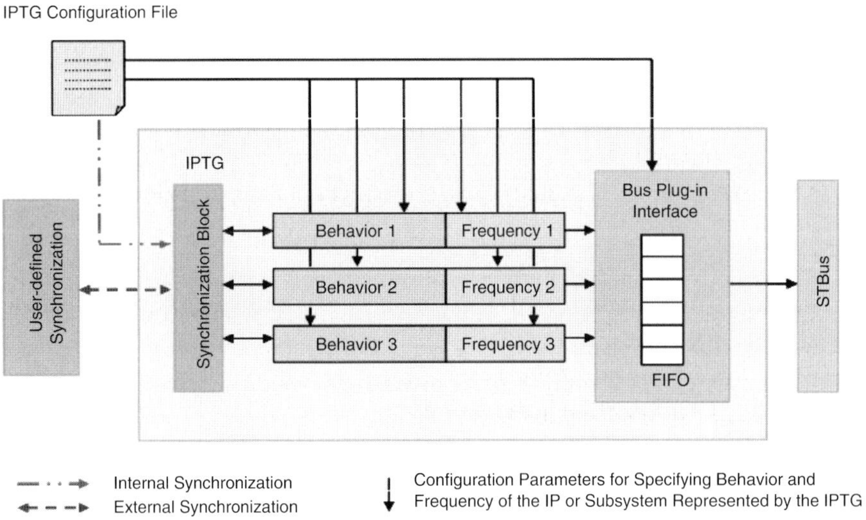

IPTG Configuration File

Figure 6-8. IP Modeling of IPTG

3.4   IPTG Configuration File

The IPTG configuration file is the key role of the IPTG methodology, which serves to model the behavior of a given IP in terms of its traffic.

Indeed, the IPTG configuration file is a text file with a set of parameters. Each parameter or more precisely, each keyword, is assigned a specific value as an argument. These values are the essential pieces of information to describe the IP traffic.

To ensure the development effectiveness and simplicity, users only need to define a subset of the parameters in the IPTG configuration file. Other parameters are kept optional. This flexibility allows not only a quick traffic definition but also a later traffic refinement during the project development.

An IPTG configuration file is divided into two sections:
1. *Header Section.*
   This section contains general description of an IP.
2. *Behavior Section.*
   This section provides specific characteristic descriptions of an IP.

Each section holds a list of keywords that are either compulsory or optional. The IPTG configuration file is written up by choosing the proper keywords and assigning them with the corresponding argument values. A particular grammar must be followed to develop both sections.

An IPTG configuration file could be manually written by architects or IP developers. The analysis tool, SysProbe, can also generate such a configuration file for a given IP. It monitors the RTL/TLM simulation of the IP and generates the corresponding IPTG file as illustrated by Figure 6-9.

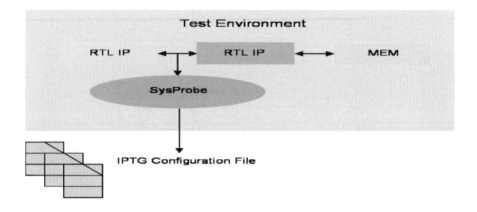

Figure 6-9. Generation of IPTG Configuration File by SysProbe

3.5 Synchronization

In order to manage synchronization issues, the IPTG incorporates a mechanism where the IPTG behavior and the bus interface FIFO are synchronized to get all the possible traffic combinations. Two approaches are distinguished for implementing synchronization in IPTG methodology:
- Configuration file-controlled synchronization;
- User-defined synchronization.

Recall that there are two synchronization blocks depicted in Figure 6-8. The synchronization block residing within the IPTG is controlled by a configuration file while the user-defined synchronization block is external to the IPTG.

3.5.1 Configuration File-Controlled Synchronization

The IPTG configuration file is extended to include the information of timing constraints specific for each behavior of an IPTG. Such information is characterized by a set of configurable parameters, which will be allocated to the current transfer in a system. Controlled by these timing parameters of the configuration file, the synchronization of an IPTG based platform can be adequately respected. Two approaches can be distinguished in handling the synchronization controlled by the IPTG configuration file:
- Linked Synchronization.
- Event-driven Synchronization.

The linked synchronization is intended for "linking" certain IPTGs together according to a set of predefined configuration rules. There are links based upon various criteria such as:

a) Process-based synchronization: A given IP is modeled by a set of processes (also called behavior). This mode assures synchronizing the different processes within a given IPTG. It is also a synchronization mode used to release time synchronization between processes coming from different IPTG.

b) FIFO-based synchronization: As illustrated in Figure 6-8, an IPTG can include a FIFO. Thus, several basic synchronizations have been developed to guarantee the synchronization of such FIFOs between different IPTGs. This feature is normally used to represent an IP that includes several bus ports. Each port is representing by an IPTG. Through such mechanism, the IP can be created by grouping all these IPTGs together.

c) Block-based synchronization: a set of traffic generators consuming data on a block-based policy; they are synchronized according to the end of each block.

d) Others: other synchronization policies are available but will not be described here.

Essentially, the linking synchronization coordinates the synchronization between all the processes within an IPTG parameterized by the IPTG configuration file. It also manages the synchronization between different linked IPTGs that correspond to the same IP. Note that these are both implementations for the "internal" synchronization of the IPTG blocks that represent the same IP. The configuration file is responsible for coordinating the different parts of the IP traffic. It is in charge of starting and stopping different behavior pieces that correspond to that IP. The main behavioral attributes parameterized in the configuration file for this purpose include:

1. random behavior succession;
2. randomization with increments or basic constraints;
3. single simulation for each behavior, i.e. no synchronization;
4. FIFO synchronization among different IPTGs.

On the other hand, the event-driven synchronization implements the system synchronization by coordinating the different IPTGs that correspond to the different IPs based upon some event-driven conditions. Such synchronization mechanism is directly included in the traffic definition of the IPTG configuration file.

Using the linked synchronization helps to obtain groups of IPTGs that represent the IPs with several bus interfaces. However, a much more complex synchronization mechanism is needed to represent the real system synchronization between these IPs. For this reason, an event-driven synchronization mechanism is required.

One of the constraints to implement the event-driven synchronization was the lack of the ability to change the synchronization mode without recompiling the system synchronization policy. To solve this problem, such functionalities are directly embedded into the configuration file of the IPTG.

During the creation of the IPTG configuration file, a synchronization keyword (e.g. GEN and WAIT with an event name) can be embedded in each process. If the system synchronization is enabled, then the overall synchronization common to all IPTG will take care about these event during the runtime.

By bringing together both the linked and event-driven methods, the configuration file-controlled synchronization can implement quite complete but rather basic system synchronization. This approach involves all the IPTGs instantiated in a SoC architecture platform to create an overall traffic of the system. It works well if all the major synchronization aspects are independent of the routing system.

3.5.2   User-defined Synchronization

The user-defined synchronization is an alternative of refining the system synchronization of the SoC architecture platform. This mechanism is implemented in the form of an "external" block where the IP behavior or process is programmed using several IPTG-specific C++ APIs. The event occurrences related to the synchronization issues are managed by these APIs. To do so, users simply need to develop single or multiple control blocks to control the IP behavior. SystemC is strongly recommended as the programming language for this purpose because it offers the built-in synchronization blocks.

Although it allows users to fully program the desired synchronization, the user-define synchronization necessitates a good command of SystemC from the SoC architect and hence induces a significant coding cost. Furthermore, the user-defined SystemC block cannot be overloaded during the simulation runtime. A re-compilation is therefore unavoidable to adapt for this change.

Given the time and effort expenses in programming the user-defined synchronization block, the untimed TLM SoC platform could be an interesting alternative. Since the untimed TLM platform is indeed a fully functional platform with the system synchronization implemented within, it can thus be reused as some sort of "timing agent" to help defining and describing the IP traffic on the platform. Based on such "functional" descriptions, the corresponding IPTG configuration files are prepared by splitting the different behavior pieces according to the "functional" synchronization. These descriptions are then connected to the matching untimed TLM models on the untimed TLM platform in order to build a "timed" TLM platform. As the untimed TLM platform implements very complete system synchronization, the resultant IPTGs manage to cover the most advanced parts of the IPTG synchronization. A rather comprehensive study of the SoC architecture can be realized through the management of the overall synchronization and data dependency by this method.

3.6   IPTG Simulation Report

The IPTG generates a simulation report at the end of each simulation. If there are multiple IPTGs, a single simulation report is generated for all of them. Two key roles of the IPTG simulation report are explained hereafter:

- *Verification of Expected Traffic*
  The resultant traffic from a simulation will be compared to the expected traffic as described in the IPTG configuration file for verification. If there are any violations of the expected traffic, the simulation report will list them out as warnings. The warnings will be shown at different levels according to the degree of severity. The type of violation will be listed as well, for instance, non-achieved bandwidth.

- *Tracing Effectiveness of FIFO*
  The FIFO in an IPTG bus plug-in interface is traced by the simulation report to study its effectiveness. First, a Value Change Dump (VCD) file is traced. The VCD file contains the information of the FIFO traced against time during the simulation. Second, a set of general statistical information is computed for the FIFO throughout the simulation, for instance, the maximum/minimum value of the FIFO. Figure 6-10 shows a screen snapshot of the FIFO traffic effectiveness analyzed by the tool of Cadence SimVision. Here, users can have a direct understanding of the generated traffic with transactions and of the FIFO evolution with analog signals.

The ultimate goal of producing an IPTG simulation report is to help the platform architects to observe, verify, and eventually optimize the effectiveness of a system.

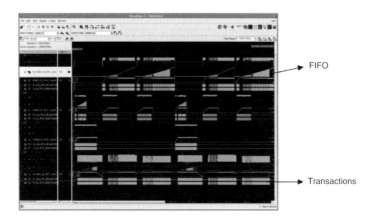

Figure 6-10. Studying Effectiveness of IPTG FIFO using Cadence SimVision

4.   ISS INTEGRATION

An Instruction Set Simulator (ISS) is often required to complete the architecture analysis of a SoC platform. Considering the complexity growth of the current SoC design, the use of micro-processors has become compulsory in most of the SoC design.

By using a timed TLM wrapper and the BFM library, the ISS can be integrated into a SoC platform at the relevant level of the architecture study. The ISS is utilized for three purposes (which will be detailed in this section):

1. COM programming;
2. interrupt analysis;
3. traffic generation.

Contradictory enough, the pitfall of using the ISS is actually driving the complexity of the SoC platform much higher. The dependency on the ISS core and the associated tool-chain are additional aspects to deal with. Sometimes, the ISS could be the bottleneck of the simulation speed unless the architecture exploration is conducted at the cycle accurate level. The ISS will however become less accurate if the architecture analysis is performed at the cycle accurate level.

Considering the irremediable tendency of using the ISS in the current SoC architecture analysis, this section will briefly discuss the three main purposes of integrating the ISS in a SoC platform.

4.1   ISS for COM Programming

Most of the communication models (COM) and memory controllers of a SoC platform require appropriate programming to assure the optimal system performance. The system micro-processor is frequently held accountable for this important task.

The IPTG can be used easily to program all the required registers of the hardware IPs for this purpose. However, this method cannot guarantee the same programming of the COM for the architecture validation and for the real software delivery. This is the main reason why the ISS is still necessary in running the SoC simulation. Therefore, SoC architects have to provide the routines to configure the COM and other critical architectural components.

Another reason to include the ISS in the SoC simulation is the potential need for updating the COM arbitration dynamically. The routines of the COM arbitration may occur upon some interrupts. Unless the whole system synchronization mechanism is successfully implemented by the IPTG, such dynamic configuration can only be achieved by applying the ISS.

4.2   ISS for Interrupt Analysis

The second typical purpose of using the ISS in a SoC architectural platform is to validate the correct execution of interrupts based on the real-time constraints.

The SoC architectural platform tailored for this purpose focuses on the ISS and the peripherals that generate interrupts. Other IPs (in the form of IPTGs) are included on the platform only for generating the noise on the interconnect and memory controllers, which assures the execution of the interrupt codes according to real traffic constraints.

The analysis based on the noise generation serves as a preliminary study of the platform interrupt and traffic. A more advanced study can be carried out by using the IPTG of all IPs involved in the platform to generate a real system-level traffic.

4.3   ISS for Traffic Generation

To better analyze a system, the SoC architectural platform should consider the traffic due to the code fetching and the cache filling. In addition, the architectural platform should also take into account the functionalities executed by the system processor core such as the MP3 treatment in a multimedia platform.

Preferably, the ISS is used to execute the code to get the real traffic for a given application. To simplify the simulation platform, however, the ISS can be replaced by an IPTG to simulate the cache refill accesses. Excluding the ISS will certainly eliminate the dependency on the ISS-specific tool-suite and debugger. The IPTG replaces the ISS by providing a generic trace that includes all of the cache refill accesses.

Before replacing the ISS by the IPTG, a simple platform consisting of the ISS and a memory is constructed to run the code. A monitoring tool is used to probe the simulation traces to create the according traffic file. This traffic file will be re-injected into the substituting IPTG. Then, a new configuration file will be created for that IPTG so that the IPTG can replace the ISS in the platform for any simulation.

5.   GETTING READY ARCHITECTURE PLATFORM

This section describes briefly of how to get ready a SoC architecture platform, covering the generic SoC architecture platform, communication model (COM), memory structure model (MEM), and the accuracy trade-off.

5.1   Generic SoC Architecture Platform

Speaking of the SoC performance analysis, the SoC platform itself would be the first thing to come across one's mind. A SoC platform is typically composed of several model blocks aimed for different purposes, for instance, the communication model (COM), memory structure model (MEM), IPTGs and other IP models.

All of these blocks could be modeled at any of the three different levels of abstraction: timed TLM, BCA or RTL. These model blocks could coexist in the same SoC platform though they might be modeled at the different levels of abstraction. Bridges are used to enable the communication among these blocks. Figure 6-11 gives a better picture of a SoC platform using the IPTG methodology to perform the architecture analysis.

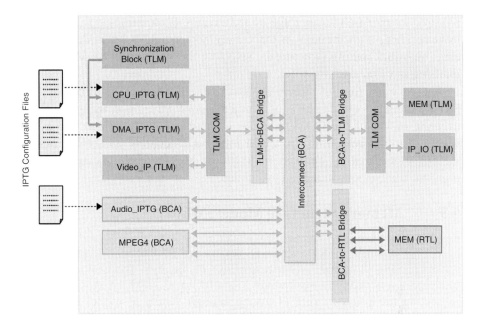

Figure 6-11. Example of Generic IPTG Platform

5.2 Communication Model (COM)

The COM is the structural backbone of the SoC platform intended for defining the communication micro-architecture features such as topology, arbitration, and FIFO size. This communication backbone can be modeled at any abstraction levels of timed TLM, BCA, or RTL, to embrace the associated communication protocol of the SoC platform.

Considering the exponential growth of SoC design today, the architecture of a typical SoC platform can easily involve around fifty initiators and tens of targets. The results of such platforms could be undesirable. Hundreds of incorrect or non-optimized routing systems may be produced along with thousands of signals holding very different programming arbitrations. For this reason, the very powerful analysis tool becomes a must in the current SoC architecture analysis.

According to the requirements of simulation accuracy and speed, a timed TLM or cycle accurate routing system is used in a given SoC project. The trade-off between the different abstraction levels for the routing system is on the account of SoC architects. Of course, the final choice is certainly dependent upon the model availability.

A timed TLM simulation aims at the early SoC architecture exploration. In this analysis phase, a very high number of simulation iteration loop is required to increase the coverage of architecture exploration up to the whole system. Then, with the known inaccuracy percentage, it helps designers to draw an initial routing structure by selecting the best suited COM type and platform architecture.

To carry out the SoC micro-architecture validation or optimization, the COM parameters need to be programmed accordingly. Thus, using the cycle accurate model of the COM becomes compulsory in this phase. To avoid wasting time in re-coding the COM into cycle accurate SystemC models, various tools such as Tenison Vtoc [7] or Mentor H2C [8] are used to translate HDL blocks into SystemC codes.

5.3   Memory Structure Model (MEM)

The MEM is a collective name designated for all models representing the memory controllers and memory modules in the SoC platform. It can be modeled at any different abstraction levels of timed TLM, BCA, or RTL, by respecting the common rule of giving enough details to model the access latency accurately.

To perform the SoC architecture analysis correctly, the TLM MEM must be configured from an ASCII file extracted from the memory specification or RTL simulation. The reason of configuring the TLM MEM is to model the "real" timing of accesses. The delay induced inside the MEM is computed based on several parameters such as previously-accessed address, type, current access, etc.

The memory is often the bottleneck of a SoC due to the memory contention. For this reason, it is recommended to model the MEM at its fullest possible accuracy. This model can be the cycle accurate SystemC model translated from HDL. It can also be the non-functional but cycle accurate blocks, which does not respect the data consistency but the cycle accuracy of transfers. This is indeed a cycle accurate memory controller without implementing the functionality of memory accesses.

5.4   Accuracy Trade-off

The proposed methodology can ensure the compatibility of analysis at different levels of abstraction. Nevertheless, this method is suggested as a complementary solution to the spreadsheet study.

As illustrated in Figure 6-12, several studies are executed according to the accuracy requirement during a project life. However, as an incremental method, the analysis is always refined. Starting from a spreadsheet study, the HW/SW partitioning as well as the basic COM and MEM choices are realized. The spreadsheets required during this initial step are reused to program the IPTG configuration file. According to the model available, timed TLM or cycle accurate simulations can finally be executed.

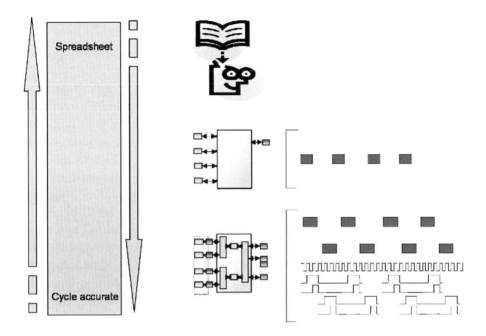

Figure 6-12. Accuracy Trade-off

6. EXAMPLE OF USING IPTG METHODOLOGY

This section provides a practical example of the SoC architectural analysis through the IPTG approach. The same methodology is used across several families of SoCs. One of this chip is the STB7100, a High Definition Low Bit-Rate Video Decoder, developed by STMicroelectronics.

6.1 Functional View of STB7100[3]

The STB7100 is the world's first single-chip Set Top Box (STB) solution supporting the High Definition H.264/AVC and VC1 specifications, which are poised to enable the next generation of high quality consumer video

---

[3] The information in this section is extracted from the website of STMicroelectronics at http://www.st.com.

systems and broadcast services. It also supports the H.264/AVC advanced video decoding standard, Microsoft's VC1 standard and high definition MPEG-2. The STB7100 can be used in:

- cable, satellite, terrestrial and IP set-top box;
- DVD in consumer and automotive.

The STB7100 demultiplexes, decrypts, decodes and outputs HD and SD video streams with associated multi-channel audio. A dual display compositor provides mixing of graphics and video with independent composition for TV/monitor and VCR outputs. SATA and USB interfaces are provided to enable low-cost connectivity to hard-disk drives and low-cost system expansion. The functionalities of STB7100 are summarized in Figure 6-13.

The STB7100 can simultaneously decode multiple HD streams and output the resultant video to two television sets, or display picture-in-picture. Its CPU core is a high-performance 300MHz ST40, ST's 32-bit RISC family based on the SuperH™ architecture and widely used across digital consumer applications. It supports all of the current STB operating systems and middleware, with power to spare for software enhancements in the future.

The new device is based on an innovative video decoding architecture which combines hardware and software techniques to allow systems to be upgraded in the field to support new standards as they become available. For Digital Video Recorder (DVR) applications it features embedded peripheral interfaces - including serial, ATA and USB 2.0 - to allow external devices to be added easily to an STB or DVD player, either during manufacture or by the viewer, in order to provide additional functionality. Viewers increasingly use digital video recording for program time shifting. Other peripherals that could be connected to a set-top box through the USB interface include digital cameras, printers, and memory cards.

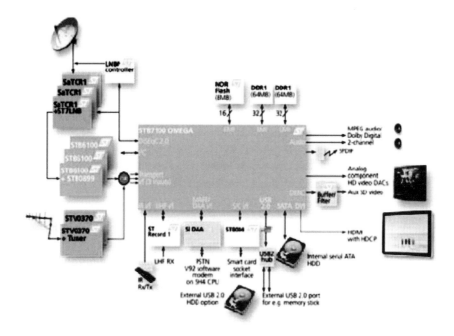

Figure 6-13. Functionalities of STB7100

6.2   Architecture Analysis of STB7100

As stated above, a typical Set Top Box or DVD SoC is built using:

- several CPUs: a host and several dedicated cores for audio and video processing;
- hardware IPs: such as hardware assists, graphic processors, and peripheral interface controllers, each of them behaving as an initiator and/or a target on the routing system;
- One or several DDR memory controllers called LMI hereafter.

The conception of the communication model for such a complex SoC starts with a spreadsheet-based analysis. The different working modes of the system are listed and characterized. For each scenario, the requirements of all the initiators are detailed then summed up in order to choose the memory buffers locations and to size the memory interfaces. Then, in order to design and validate in advance the interconnection between on-chip IPs and to configure the traffic of the IPs, the whole system is modeled in SystemC at transaction level.

This platform consists in several tens of IPTGs describing the behavior of the CPUs and the IPs' initiator side. The communication model is based on simple switches and links available both in BCA and RTL. The MEM is made of the LMIs and of basic memories modeling the IPs' target side. An example of such platforms is shown in Figure 6-14.

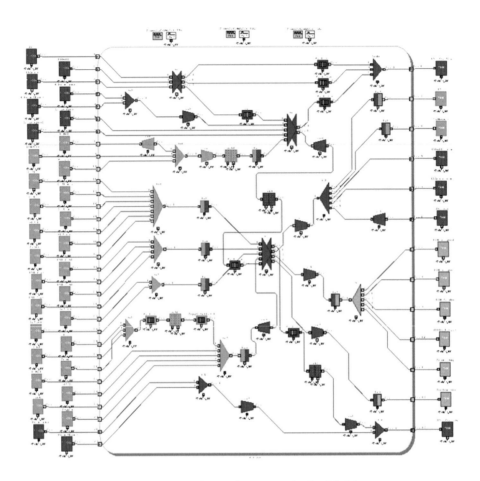

Figure 6-14. Schema of a Communication Model

The IPTGs are modeled at TLM level and the COM and basic memories at BCA level. The RTL model of the LMIs is used to obtain cycle accurate behavior for these key components, which are the bottleneck of the system. A TLM-to-BCA translator is then associated to each IPTG, and a BCA-to-RTL translator to each LMI port.

The CPUs are, in a second step, replaced by their associated ISS. The CPUs are configured in traffic modeling mode. The host is in charge of the communication model and the memory controllers.

The other IPTGs are set in IP modeling mode. An important feature of the IPTG is that it enables to model the dependencies between plugs of the same IP, thanks to the synchronization mechanism. Consider an IP that works from memory to memory, the pipeline is stopped whenever a write plug is full or a read plug is empty.

The analysis of a simulation performed on such a platform is straightforward because the IPTG FIFO level is monitored. Any over/underflow in a real-time IP is flagged and the percentage of pipeline stopped time in decoder IPs is reported.

For each working mode of the IPs, a set of IPTG configuration files is defined to model the worst case in terms of bandwidth consumption. Then, scenarios of the spreadsheet analysis are reproduced, gathering the IPTG configuration files of all the IPS, and a simulation is run for a portion of an image.

When the performances are not met, the SysProbe transaction debugger allows to observe directly internal nodes of the communication model and to analyze the root cause of the performance drop off. A side advantage of this approach is that the verification of the communication model's RTL can be done in the SystemC environment, using meaningful scenarios.

7. CONCLUSION

The methodology proposed in this chapter enables outlining a plug-and-play architecture environment based on the platform assembly that requires no new language learning.

The IPTG approach is put forward as a complementary solution to the conventional architecture analysis on paper, which takes into account the inadequacy of simulating a real system's scenario at an accurate level. By adopting this method, IP designers create directly the IP configuration file that will be reused across various projects with high flexibility to update various products. In brief, the very rewarding result of this methodology is an analysis environment that is powerful yet easy-to-maintain.

REFERENCES

[1]   OCP Specification, Available on the OCP Website: http://www.ocpip.org

[2]   STBus Functional Specifications, Available on STMicroelectronics Public Support Website: http://www.stmcu.com/inchtml-pages-STBus_intro.html, April 2003.

[3]   ARM AMBA 3.0 Specification, Available on ARM Website: http://www.arm.com

[4] Cadence Incisive (SimVision), Information available on Cadence website: http:///www.cadence.com

[5] Novas Verdi, Information available on Novas website: http://www.novas.com

[6] Cadence TxE, Information available on Cadence website: http//www.cadence.com

[7] Tenison Vtoc, Information available on Tenison website: http://www.tenison.com

[8] Mentor Graphics H2C, Information available on Mentor Graphics website: http://www.mentor.com

Chapter 7

DESIGN AUTOMATION
Integrating TLM in SoC Design Flow

Christophe Amerijckx[1], Stephane Guenot[2], Amine Kerkeni[3], Serge Hustin[1]

STMicroelectronics Belgium[1]; STMicroelectronics France[2]; STMicroelectronics Tunisia[3]

Abstract: Although the TLM development and usage only require a C++ development environment and a SystemC library, design automation is the key to integrating TLM in the SoC design flow for further reaping the design productivity and quality rewards brought by TLM. This chapter explains how TLM has been integrated in the design flow at STMicroelectronics both by extending the SPIRIT XML packaging standard to support TLM and by developing the tools needed to integrate TLM in the flow.

Key words: SPIRIT; SystemC TLM; SoC; design flow; design automation; XML; data model and schema; platform assembly; meta-level; content-level; configurator; generator; netlister; IP packaging; platform generation.

1.   INTRODUCTION

The minimum tool and library requirements for the TLM methodology are simply a C++ development environment and SystemC classes. A flawless integration of the TLM methodology into the SoC design flow, however, entails further tools and libraries implementations. This chapter describes at length the necessary accompanying implementations to make the best use of the TLM methodology in the SoC design cycle.

This goal is attainable through establishing and enforcing a standard *automation* strategy to integrate the TLM methodology into the essential phases of the SoC design flow. The TLM assembly should therefore adopt an automation approach ranging from the design database to editor, configurator, checker, and netlister.

F. Ghenassia (ed.), Transaction Level Modeling with SystemC, 241-266.

From a given design database, TLM components are instantiated, configured, and interconnected by a platform editor. Configurators and checkers are subsequently employed to propagate the redundant information through the design description, and to verify the design integrity. Lastly, netlisters project the design description into its different targets, covering not only TLM but also verification, software, and hardware emulation.

To support all of the automation tools above, a *common format* must be adopted to store the design information as well as to package the design components.

The remainder of the chapter provides a detailed description of the component and design representations, followed by an in-depth explanation of the automation tools, and finally an illustration of their applications on a real design.

2.   DESCRIPTION OF DESIGN AUTOMATION

2.1   Introduction

With the advent of the explosive nanotechnology era, the design of the System-on-Chip (SoC) is getting increasingly complex without the help of efficient tools. The SPIRIT[1] Consortium [1] has developed a standard mechanism for describing and handling IPs, with the aim of accelerating large-scale SoC designs through automated configuration and integration of the designs [2].

SPIRIT provides an eXtensible Markup Language (XML) schema to describe components and designs. Rules and regulations are also imposed by SPIRIT for implementing the user interfaces of automation tools, such as generators and configurators, to handle SPIRIT compliant components or designs. The SPIRIT design environment is clearly illustrated in Figure 7-1.

For each component or IP in a given SoC design, SPIRIT defines a specific XML file containing the metadata that will be used by a SPIRIT compliant design tool. The content of such component XML file is defined in the SPIRIT schema for the following aspects:

1. *Bus interfaces available for a component.*
   This description allows the automation of connecting a component to different components of the same interfaces. The bus interface here refers

---

[1]   Structure for Packaging, Integrating and Re-using IP within Tool-flows.

to a bus definition that specifies the bus Vendor Library Name Version (VLNV).

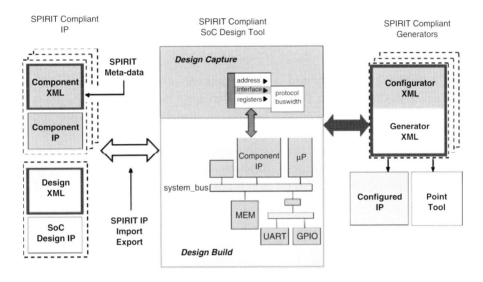

Figure 7-1. SPIRIT Design Environment

2. *Different views available for a component.*
   The descriptions of the different views available for a component are essential in determining the abstraction levels for that component within a given design. Each view refers to a file set that holds all of the files specific to that view.

3. *Memory map and remap information.*
   Providing the memory map and remap information of a component is intended for specifying the different registers available on the slave interface of that component.

4. *Address space.*
   The address space of a component must be described because it defines the logical space accessible by the master interface of the component.

5. *Hierarchy information.*
   If a component is hierarchical, the information of the different associated component instances must be provided along with their interconnections.

6. *Configuration parameters.*
   The different parameters available for configuring a component are described, for example, the size of a RAM, the number of master/slave of a bus, etc.

7. *File sets of different component views.*
   A list of the file sets that specifies the various files used by each view of a component has to be provided.

A SPIRIT compliant tool uses the content of the SPIRIT metadata to automate the SoC design through the followings:

1. instantiation of components in a given design followed by prompting users for the view selection;
2. automatic connection of components depending on their bus interfaces;
3. prompting users for the configuration of parameters;
4. launching SPIRIT compliant generators.

The top-level structure of a SoC design is specified by its different component instances along with their connection at the bus interface level via interconnections, or at the point-to-point level via ad hoc connections for all the signals not belonging to a bus. The ad hoc connections are of course present only at the RTL level.

A SPIRIT generator can be launched from a design tool to accomplish those tasks that are not managed by the design tool, for instance, netlist generation, configuration, compliancy, consistency checking, clock-tree generation, etc.

The coming sections further describe how the design automation is achieved in line with SPIRIT. The initial version, i.e. SPIRIT V1.0, focuses only on the RTL hardware view. We shall therefore start our discussion from this single-view approach. The next major release, i.e. SPIRIT V2.0, will allow the co-existence of the multiple-view such as untimed TLM, timed TLM, bus-cycle accurate (BCA), and RTL. Our discussion will highlight the ST Microelectronics view on how such multiple-view approach is completed.

2.2   Single-View Component Structure

The single-view (i.e. RTL) component structure pertaining to the SPIRIT V1.0 standard is detailed hereafter.

2.2.1   Bus Definition

The specification of a bus is stated by a bus definition. The bus definition is identified by its VLNV. Another important piece of information stated within the bus definition is the maximum number of masters and slaves that a bus can hold. In addition, the bus definition contains information specific to the RTL view. This includes the collection of signals that belong to the bus, and of the constraints to be applied to these signals such as directions on master/slave and default values.

Indeed, the bus definition is strongly analogous to VHDL in the sense that the bus definition is a kind of VHDL record type whereby a bundle of signals belonging to the same group can be defined.

2.2.2 Bus Interface

Different components are interconnected by the bus interfaces defined for each component. Each bus interface within a component is designated a specific name. Two interfaces can be connected together if:

1. they are of the same type as identified by the VLNV of the bus definition;
2. they have matching interface natures (e.g. master, slave).

For this reason, a bus interface must be specified in terms of its type and nature as illustrated by the following example.

Example of Bus Interface Definition

```
<spirit:busInterface>
   <spirit:name>ambaAPB</spirit:name>
   <spirit:busType       spirit:vendor="AMBA"        spirit:library="AMBA"
spirit:name="APB"
      spirit:version="v1.0"/>
   <spirit:slave>
      <spirit:memoryMapRef spirit:memoryMapRef="ambaAPB"/>
   </spirit:slave>
   <spirit:signalMap>
      <spirit:signalName spirit:busSignal="PSELx">psel</spirit:signalName>
      <spirit:signalName spirit:busSignal="PENABLE">penable</spirit:signalName>
      <spirit:signalName spirit:busSignal="PADDR">paddr</spirit:signalName>
      <spirit:signalName spirit:busSignal="PWRITE">pwrite</spirit:signalName>
      <spirit:signalName spirit:busSignal="PWDATA">pwdata</spirit:signalName>
      <spirit:signalName spirit:busSignal="PRDATA">prdata</spirit:signalName>
   </spirit:signalMap>
</spirit:busInterface>
```

At the RTL level, an interface represents a group of signals. The signal names defined in the bus definition may not always match the signal names defined in the design. Therefore, a section of *signal mapping* is included in the bus interface for indicating the relationship between these two signals. As shown in the example above, the signals defined in the bus definition (i.e.

logical signal names) are mapped to those defined in the design (i.e. physical signal names).

Similar to the analogy of the bus definition with VHDL, the definition of a bus interface looks like a signal definition whose type would be the one defined for the bus definition using a record.

2.2.3   View Specification

In SPIRIT, the *view* of a component represents a level of abstraction or an implementation of a particular component. The specification of a single-level view is quite straightforward. First of all, a given view must be designated a specific name in order to distinguish it from other views. An environment identifier further specifies the environment that can be used by the given view, for instance, simulation and synthesis environments. The language applicable in the given view such as VHDL, Verilog or SystemC must be stated as well. Lastly, a reference to a file set listing all the files delivered with that particular view is specified, along with the HDL-specific model names. Quoted below is an example of the specification for a component with the RTL view.

Example of the RTL View Specification

```
<spirit:view>
   <spirit:name>RTL</spirit:name>
   <spirit:envIdentifier>Simulation</spirit:envIdentifier>
   <spirit:envIdentifier>Synthesis</spirit:envIdentifier>
   <spirit:language>vhdl</spirit:language>
   <spirit:modelName>leon2_Uart(struct)</spirit:modelName>
   <spirit:fileSetRef>vhdlSource</spirit:fileSetRef>
</spirit:view>
```

The SPIRIT V1.0 standard assumes that each view contains all of the defined bus interfaces, implying that bus interfaces are neither optional nor specific to a particular view.

2.3   Multiple-View Component Structure

The multiple-view (i.e. untimed/timed TLM, BCA, and RTL) component structure proposed by ST Microelectronics for the SPIRIT V2.0 standard is detailed hereafter.

2.3.1   Bus Interface

To support multiple hardware views, SPIRIT V2.0 will add a new element in the bus interface schema to specify the name of the available abstraction level for the bus interface. The specification of signals is shifted to the abstraction level of RTL. As an example, a given bus interface may support all of the abstraction levels available for a design while another bus interface may only support RTL and TLM abstractions.

One may claim that such added feature is duplicated information because the abstraction level of the bus interface should be the same as the one of the component. However, a specific abstraction level of a given component may not always have the same name as the corresponding abstraction level in the bus definition. The naming convention helps to enforce a mapping of the abstraction levels between the component and the bus definition. For example, the abstraction of the timed transaction level could be named as "TLM-timed" in the bus definition while it could be named as "PVT" in the component.

2.3.2 View Specification

The view specification of a component with multiple abstraction levels remains the same as the view specification of the single-view component. The only difference is that meticulous care must be given to handle bus interfaces. In multiple-view components, certain bus interfaces may be present in a particular view but absent in another view. A typical example of such is the test interface that exists at the RTL but not at the untimed TLM, as depicted in Figure 7-2. Note that this a distinct difference from the SPIRIT V1.0 where all bus interfaces are defined for every component view.

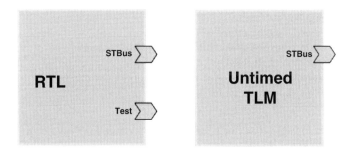

Figure 7-2. Bus Interfaces of a Component with Two Abstraction Levels

To resolve such problems, each component view must include the list of bus interfaces available in a design. This feature facilitates the SPIRIT compliant tool to retrieve the available bus interfaces for each abstraction level of the component.

Consider the example of a component with two abstraction levels, RTL and untimed TLM, as illustrated in Figure 7-2. At the RTL level, the component has two bus interfaces. The first is a STBus level-2 interface while the second is a test interface. At the untimed TLM level, a STBus level-2 interface is the only present interface. The corresponding XML code of this multiple-view component is provided hereafter. Note that *busInterfaceNameRef* refers to the name of the bus interfaces defined in the bus interface section of the component XML metadata file.

Example of the XML Code for a Multiple-View Component

```
<spirit:views>
   <spirit:view>
      <spirit:name>RTL</spirit:name>
      ...
      <spirit:interfaceList>
         <spirit:busInterfaceRef>
            <spirit:name>STBus<spirit:name>
            <spirit:busAbstraction>RTL</spirit:busAbstraction>
         </spirit:busInterfaceRef>
         <spirit:busInterfaceRef>
            <spirit:name>test<spirit:name>
            <spirit:busAbstraction>RTL</spirit:busAbstraction>
         </spirit:busInterfaceNameRef>
      </spirit:interfaceList>
   </spirit:view>
   <spirit:view>
      <spirit:name>PV</spirit:name>
      ...
      <spirit:interfaceList>
         <spirit:busInterfaceRef>
            <spirit:name>STBus<spirit:name>
            <spirit:busAbstraction>PV</spirit:busAbstraction>
         </spirit:busInterfaceRef>
      </spirit:interfaceList>
   </spirit:view>
</spirit:views>
```

2.4   Design Structure

In our context, a *design* is the representation of the top-level structure for a given SoC platform. A SPIRIT compliant design contains all of the instances and connections that form a SoC.

There are three compulsory sections of a SPIRIT compliant design:

1. VLNV of the top-level design;
2. different component instances instantiated at the top-level;

3. connections between component instances at the bus interface level and point-to-point level.

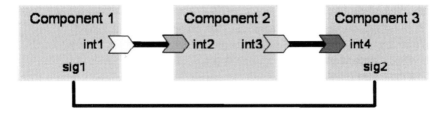

Figure 7-3. A Design with Three Components

Figure 7-3 demonstrates a simple design made of three components interconnected by:
a) bus interfaces from int1 to int2, and from int3 to int4;
b) point-to-point connection between sig1 and sig2.

The corresponding XML description of this component is provided hereafter. The description starts with the VLNV of the design, giving information on the design vendor, library, name, and version. The next section, *<spirit: componentInstances>*, lists all of the component instances. Each of these instances holds an instance name and a reference to access the component library identified by its VLNV. Following this is the section of *<spirit: interconnections>*, which provides the information on the interconnections of bus interfaces. For every component involved in the interconnection, there are two values required by this section: the name of the component instance and the name of the bus interface on that component. The last section, *<spirit: adHocConnections>*, gives the specification of the point-to-point connections between signals.

Example of XML Design Representation

```
<spirit:design>
   <spirit:vendor>ST</spirit:vendor>
   <spirit:library>Example</spirit:library>
   <spirit:name>simple_design<spirit:name>
   <spirit:version>1.0</spirit:design>
   <spirit:componentInstances>
      <spirit:componentInstance>
         <spirit:name>component1</spirit:name>
         <spirit:componentRef spirit:vendor="ST" spirit:library="Processor"
            spirit:name="proc1"  spirit:version="1.0"/>
      </spirit:componentInstance>
      <spirit:componentInstance>
```

```
        <spirit:name>component2</spirit:name>
        <spirit:componentRef spirit:vendor="ST" spirit:library="Bus"
            spirit:name="bus1" spirit:version="1.0"/>
    </spirit:componentInstance>
    <spirit:componentInstance>
        <spirit:name>component3</spirit:name>
        <spirit:componentRef spirit:vendor="ST" spirit:library="Peripherals"
            spirit:name="uart" spirit:version="1.0"/>
    </spirit:componentInstance>
</spirit:componentInstances>
<spirit:interconnections>
    <spirit:interconnection
        spirit:component1Ref="component1" spirit:busInterface1Ref="int1"
        spirit:component2Ref="component2" spirit:busInterface2Ref="int2"/>
    <spirit:interconnection
        spirit:component1Ref="component2" spirit:busInterface1Ref="int3"
        spirit:component2Ref="component3" spirit:busInterface2Ref="int4"/>
</spirit:interconnections>
<spirit:adHocConnections>
    <spirit:adHocConnection>
        <spirit:pinReference componentRef="component1" spirit:signalRef="sig1"/>
        <spirit:pinReference componentRef="component3" spirit:signalRef="sig2"/>
    <spirit:adHocConnection>
</spirit:adHocConnections>
</spirit:design>
```

No information regarding the abstraction level of the design components is provided in the XML description. This aspect will be handled in a separate file. One way to handle this is providing a default file that gives the preferred list of view for each component instance. The tool will check through the default list for the *first* available view. Once found, that view will be accepted by the tool. The default rule can be overwritten by another if the designer would like to change the selected view. This is quite a common routine in the design process where all of the components are initially at untimed TLM level, and then some components are gradually changed into RTL. Such manipulation continues until a complete RTL platform is obtained at the end.

Since no information is given for the abstraction level in the design, the role of generators is vital. The netlister must verify the abstraction level for each component to assure the right connections. Two component interfaces of the same abstraction level can obviously be connected directly. If their abstraction levels are different, the netlister will have to insert a transactor - or BFM - between the two interfaces. This is typically the case where a component is of untimed TLM level while another is of RTL level. In such cases, the role of the transactor will be converting transactions into signals.

3. AUTOMATION TOOLS

3.1 The Need for Platform Assembly Automation

The Integrated Circuits (IC) industry has been growing exponentially for a few decades. ICs are no more simple chips with a few components for a specific functionality but System-on-Chip (SoC) with multi-millions gates for a whole system. The current SoC industry must treat every step in the SoC design flow as much as possible at the platform level, where the system behavior is studied and managed through the communication between platform IPs.

Once a given SoC design is simulated at TLM level, it will have to go for the RTL simulation where all cycle-accurate signals must be connected. Not only is the RTL simulation a time- and effort-consuming job, it is also highly error-prone.

Consequently, the need for the platform assembly automation has become more and more critical nowadays. The concept of using the data model in the platform assembly operation is relatively clear for its users. However, this concept is not really implemented in the industry today because the optimization of the high-level platform simulation does not allow treating such data models (which represent very often the transactions between platform components).

With the ever-rising SoC complexity, the need for automating the SoC platform assembly should no longer be compensated by such optimization. The most noticeable advantages include reduced error probability and immediate productivity enhancement. In addition, any modifications on an existing platform description can save SoC developers a lot of time. The SoC flow automation needs to tackle two areas:
1. automation of standard tasks by using SPIRIT compliant generators;
2. providing SPIRIT compliant tools for tasks that require user inputs.

3.2 Foundation of Flow Automation

All kinds of industrial activities follow specific procedures that can be described as a *flow*. Without exception, the SoC industry must obey this rule as well. Various methodologies were developed to assist SoC developers in formalizing the SoC design flow. Formalizing the SoC design flow means identifying tasks that can be fully automated and those that require user input.

Note that the majority of the automated tasks can only be executed upon the input of the necessary information from users. To create such formalization, two parts of the flow must be distinguished:

1. *Structural Part.* This part consists of user-dependent data, environment-dependent data, configuration parameters, data constraints, and any other data that are collected as input for a set of automation tools.

2. *Functional Part.* The functional part is the data treatment in a flow.

Both parts can be formalized into the format of Unified Modeling Language (UML), which is a methodology describing the flow of any activities.

Since long, there exists an absolute formalized data model in the algebra called Relational Data Base (RDB). This universal data model is quite simple but very efficient. Through a data model diagram, the RDB represents the formal structure and logical relation of the input data for a given activity flow. Otherwise stated, the RDB describes and outlines the structure of data.

Figure 7-4 illustrates an example of the data model with its data classes, data fields, and the data hierarchy. This diagram is an interpretation of the SPIRIT data model that covers only the description of hardware connections. It shows the non-exhaustive examples of data classes and data fields, and cardinal relations between the classes.

There are loads of different methods to realize the diagram or *schema* of the RDB model. To design this schema, the data structure that will be described by the RDB model must be carefully developed, including the hierarchical and cardinal relation for the whole data structure.

Many commercial data model tools called database engines are available in the market today. These tools access to the database through the standard request language, Simple Query Language (SQL). However, these tools often require proprietary database engines to store data in the proprietary binary format. These proprietary tools have consequently made SQL a non-standard language without a universal format.

For this reason, a descriptive language should be used to handle the structural part of a flow. The XML is strongly recommended as the best solution for this purpose since it can hold any formalized contents. The XML documents can be manipulated by using standard parsers and validators such as Xerces from W3C[2]. These parsers and validators check the consistency of the described elements in the XML (like what an RDB engine

---

[2]   Refer to World Wide Web Consortium (W3C) at http://www.w3.org.

will do), e.g. checking the descriptions of platform design, components/IPs, and bus specifications/definitions.

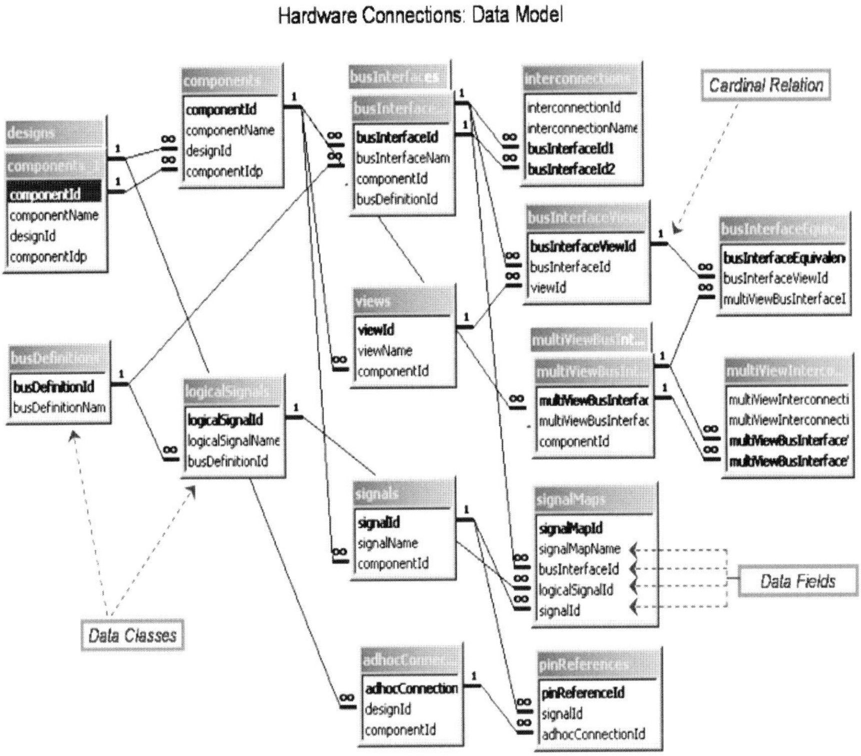

Hardware Connections: Data Model

Figure 7-4. Example of Data Model

Although the XML can handle data contents very well, the data structure must be described. Different methodologies are available to provide the technical support for describing data structures. Be it any method, a list of "containers" has to be designated to describe the data model. These containers are the data classes depicted in Figure 7-4; they are called *table* in the RDB, *class* in the object-oriented language, and *sequence* in the XML Schema Definition[3] (XSD). For each container, a set of data fields must be defined as depicted in Figure 7-4. These data fields are called *column* in the RDB, *member* in the object-oriented language, and *element* in the XSD.

---

[3] Recommended by W3C to formally describe the elements in the XML documents.

The real strength of a data structure descriptive tool lies in its ability to illustrate the relation between the container elements, rather than the description of the container itself. There are two fundamental types of such relations distinguished by their cardinality:

1. *One-to-One Relation*. A given data member can only be related to a single data member from another data class, for example, a set of parameters added to a given class.
2. *One-to-Many Relation*. A given data member can be related to one or more data members from other data classes, for example, a microelectronics component can hold several bus interfaces.

The data model exists most of the time in an implicit form in many industrial activities. The explicit implementation of the data model, on the other hand, is able to offer very extensive applications ranging from assembly tools to generator tools.

SPIRIT has adopted the XML format, a universal standard opened to the public, to describe a data model. As such, the XML specification cannot verify the relational and cardinal integrity of the data model described by the XML documents. A list of semantics rules called "grammar" or "schema" must be implemented to further describe the data model. Certain specific languages can formalize the XML semantics rules; among which, the most advanced description methodology is the XSD. The XSD schema is however limited to describe all the necessary semantics rules. For this reason, there are two principal parts in the SPIRIT standard:

1. XSD schema for describing the technical content;
2. User guide for describing the semantics rules and the design automation flow.

3.3   Strategy for Automation Tool Development

This section discusses the strategy adopted for developing the automation tool in line with the SPIRIT standard.

3.3.1   SPIRIT Meta-Level Description

Be it any methodology or description language of data model, the structure of the data model, i.e. the data classes, data fields, and the relation between data classes, must be described by formal rules. These rules are *meta-level* rules since they implement the "data model" of a data model, i.e. *meta* data model.

All of the description methodologies provide the syntax to create and update the meta-level description. In the SPIRIT standard, the meta-level descriptions for XML documents are implemented in XSD schema. The semantics rules in charge of validating the meta-level description itself are usually hard-coded in the database engine, for instance, hard-coded in Xerces for the XSD schema.

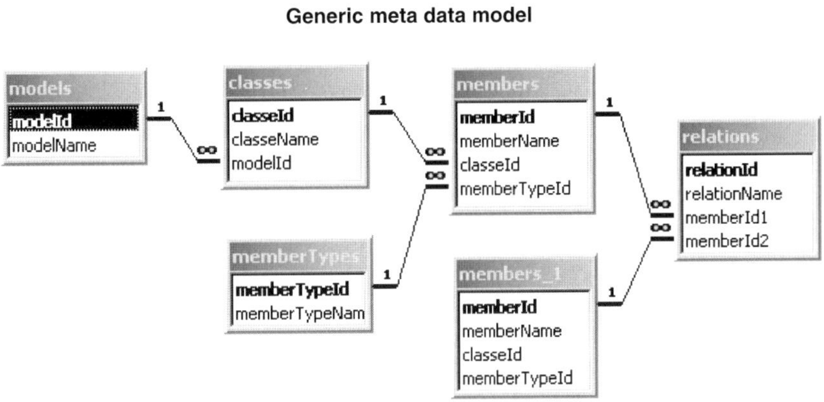

Generic meta data model

Figure 7-5. Description of Generic Meta Data Model

Figure 7-5 gives a description of the generic data model at meta-level. Note that in a SPIRIT meta data model, a list of data classes (also known as table or sequence) must be defined. Every defined data class holds a list of members (also known as element or column) with a unique type each. The cardinal relation between the members can be described as well.

3.3.2 SPIRIT Content-Level Description

Once the SPIRIT meta data model is implemented, its structure must be filled up with the necessary contents[4]. The SPIRIT XML documents are created to store the database of such contents.

An important goal of the SPIRIT platform description is to provide an input for a set of tools such as generators. Various Application-Programming Interfaces (API) can be used to implement the SPIRIT compliant tools. For any kinds of API as C++ or Java, the data must be treated first. To do so, the API must contain a list of methods to create, modify, and delete the SPIRIT objects such as bus interfaces.

---

[4] See section 3.4.2 about the editor tools for editing the contents.

The W3C consortium provides a specific API, Xerces, which is divided into two main parts:

1. *Validator and Parser*: parse XML documents and then load the XSD schema; also check the content integrity of XML documents, and fill a proprietary Xerces structure if the integrity is well respected.
2. *Document Object Model (DOM)*: a low-level API.

The DOM API provides very generic structures in the form of a simple data tree. It also provides methods to manage the data tree, i.e. methods to access the SPIRIT contents through the software. However, the DOM API does not take into account the XSD schema and thus not allowing efficient programming of SPIRIT compliant tools. A higher layer dedicated to the SPIRIT schema is therefore required as explained in the next section.

3.3.3   API Generation

As explained earlier, a higher layer API is necessary to handle the SPIRIT schema or the meta data model. A standard approach to produce this layer is writing it manually in an appropriate programming language such as C++ or Java. This manual task is always a very time- and effort-consuming job due to the huge schema size.

A more interesting approach is to make use of the SPIRIT schema to generate the SPIRIT-specific API structures, and their *exhaustive* access methods to the entire content described by the XSD schema. This approach is feasible because the meta-level description is formalized in the XSD schema, which is indeed a kind of XML documents. The standard DOM API allows loading the XSD schema for filling up the corresponding meta data models in the form of C++ meta structures. Essentially, such C++ structures are filled in the memory after being analyzed by the model builder tool as the representation of the SPIRIT meta data model.

Once this abstract representation is available, the different applications can be generated, e.g. all the translation tools from/to the SPIRIT XML format. The most important generation is the SPIRIT API source code, which represents a "snapshot" of a given SPIRIT version. This allows an automatic re-generation of API source code if the SPIRIT version is updated or when proprietary schema extensions are implemented. The choice of the target language for the generated API is independent of the language of the generator itself; thus, it can be in any language such as C++ or Java[5].

---

[5]   For the future version of SPIRIT platforms, such generations are in C++ for simulation compliance reasons.

With this generated API, SoC developers will no longer have to deal with a generic data tree structure based upon DOM but concrete SPIRIT compliant objects such as components and bus interfaces. All of the elements of the SPIRIT schema such as signal direction or signal size will be created explicitly during the API generation as the class members. These data members can be manipulated explicitly by the SPIRIT developer.

The dedicated C++ structures with access methods to the members are not the only necessary targets. The generated API must provide the methods to instantiate, update, duplicate, and remove SPIRIT objects. However, a further need is an immediate method to load the XML document into the C++ structure, and to dump this structure in a new XML document after modification. This is where a loader and a dumper are required. Figure 7-6 and 7-7 summarize the discussion of the previous three sub-sections.

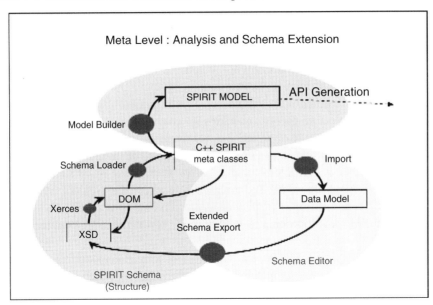

Figure 7-6. SPIRIT Meta-Level

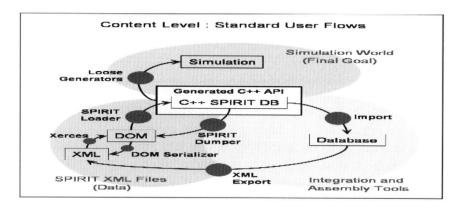

Figure 7-7. SPIRIT Content-Level

3.3.4   Development Environment and Inter-operability

The SPIRIT development environment is the "meta-level environment" for the SoC design environment, which is indeed analogous to the relationship between meta-level and content-level SPIRIT data models. The development environment provides SoC developers with the necessary tools and facilities to create a user-specific design environment.

To optimize the work of SPIRIT developers, a full SPIRIT development environment including the generated API should be made available. The generated API is nevertheless an exhaustive C++ view of SPIRIT schema without the semantics rules. As mentioned earlier, the SPIRIT semantics rules are collected in the SPIRIT user guide. To enable verifying these rules throughout the development, a higher API layer must be implemented manually[6].

This higher API facilitates the writing rules by encapsulating the low-level methods to complement the work of the generated API. Furthermore, this layer separates the generated API from the tools developed by SPIRIT developers, i.e. SPIRIT tools are independent of the generated API. As the SPIRIT version evolves, tool developers simply need to re-generate the API for the XSD schema and update the API for the semantics rules while the SPIRIT tools themselves are kept untouched.

Another fundamental goal of SPIRIT is to provide a tangible solution of flexible inter-operability among different tools provided by different EDA vendors or IP providers. SPIRIT describes the generators of the SoC design flow with standard interfaces at every flow step. The SPIRIT development

---

[6]   STMicroelectronics has developed a checker tool to verify the SPIRIT semantics rules (see section 3.4.4).

environment contains various "building bricks" such as Loose Generator Interface (LGI) and Loose Generator Change (LGC) (see section 3.4.5), which allow end users to swap from one SPIRIT compliant tool to another at any step throughout the flow.

3.4   SPIRIT Compliant Automation Tools

At the entry point of the SPIRIT design flow, the data required in the flow activities must first be imported either manually or automatically by some scripts or tools. Once entered in the SPIRIT environment, the imported data will be typed or edited by specific editors. The SPIRIT compliant automation tools will then treat the data in line with the objective of end users, which is the SoC design simulation for SoC developers.

The SPIRIT compliant automation tools are classified into five families:
1. Packager.
2. Editor.
3. Checker.
4. Configurator.
5. Generator.

3.4.1   Packager

Automated processes require the input data to be packaged or put together according to the technical specification. The SPIRIT compliant *packager* is an automation tool for packing all the input data into the XML description of microelectronics components.

If the input data exists already in a specific formal format such as FrameMaker, RTL or SystemC, a set of scripts are provided to translate such data into a SPIRIT XML document.

3.4.2   Editor

Once the imported data is translated into the XML format by the SPIRIT compliant packager, the design description requires some meta data that is necessary to enable the design configuration and automation. Such meta data is usually prepared and entered manually by SoC developers. For this reason, an editor is needed in the flow to edit and modify the SPIRIT XML database.

As an ASCI format, the XML document allows any text editors to process and package a component description. The end users, however, expect something more user-friendly than the XML format such as an editor with a Graphical User Interface (GUI).

Two types of GUI editors are available:

1. *XML Generic Editor*. An editor that edits and modifies the XML documents based on any XSD grammar, e.g. SPY.

2. *SPIRIT-Specific Editor*. A specific editor that respects the SPIRIT grammar and schema. The XSD methodology adopted by SPIRIT is consistent enough to provide automatic ways to create GUI for XML packaging. This is a tool generation that is similar to the API generation from the same representation of the SPIRIT meta data model (see section 3.3.3). The most important packaging parts, however, remain the manual optimization process.

The result of an editor is an XML document for all of the components of a design. A particular tool is needed to assemble all of these components and interconnect them to form a design or platform. Therefore, a specific editor, *platform assembler*, is created to perform this job.

The platform assembler can be in the form of GUI where users can select any components to instantiate in the platform, interconnect bus interfaces, and connect signals not belonging to any bus by ad hoc connections. In addition, the assembler tool also configures the parameters of the component instances. The result of the platform assembler is a new XML document file for the design, i.e. Design XML file.

3.4.3   Checker

The SPIRIT compliant checker is a specific tool developed for verifying the semantics rules written in the SPIRIT user guide.

The SPIRIT schema allows checking many integrity constraints in a design description. However, certain description methodologies cannot check all of the constraints needed to verify a design description.

Therefore, an applicative layer called *checker* is added on Xerces, the standard XML validator. This layer implements the semantics rules for the SPIRIT schema and the reference validity for elements from different files.

The latter task cannot be performed by the XSD since it can only treat a single file at a time. As an example, the checker must be used to verify the names of the components instantiated in a design because all of the components have their own separate XML documents. The checker can be invoked anytime at any step in the design flow.

3.4.4   Configurator

The SPIRIT compliant configurator is a tool that configures the SPIRIT data according to the design context. This configuration is based on either:

1. a template, e.g. configuration for replicating interfaces;

2. or the user input in line with the design context.

Just like the checker, the configurator can be invoked can be called repeatedly after any iteration from the Design XML.

Indeed, the configurator is an XML-to-XML tool. Given the SPIRIT input as an XML document, it is configured by the SPIRIT configurator to produce a SPIRIT output that is another XML document but with some modifications or configurations.

3.4.5   Generator

The SPIRIT compliant generator is a very important tool family. In the design flow, several generators can appear together as a *generator chain* with each targeting a specific task. For the SoC design flow, the key generator is of course the netlister (see section 3.5 for further discussions).

Typically, a generator reads a complete SPIRIT design description as the input data in the XML format. A generator chain is then created for that design. Each generator of the generator chain, for the reason of inter-operability, contains three elements of SPIRIT compliant generators:

1. *Loose Generator Interface (LGI)*.
   This sub-generator takes the design environment as the input to create an LGI file, which holds the access path to all of the XML documents of the design environment. It helps to implement a generator tool that is independent of the design environment, i.e. only the LGI will have to be modified if the access path to an XML file is changed.
2. *Function-Specific Generator*.
   This is a generator with specific tasks, e.g. the netlister to produce a netlist.
3. *Loose Generator Change (LGC)*.
   If any function-specific sub-generators make some changes in XML documents, then the LGC must write these changes into an LGC file to update the design environment.

Indeed, the generator chain is described in the XML format as a "meta-generator" that permits the SPIRIT compliant tools to perform the entire generation flow from the design description in a single shot. The meta-generator generates all of the necessary output of the design (e.g. netlist), which will be utilized by the simulation tool such as the SystemC-RTL simulator.

3.5   Netlister

The netlister is a particular type of generator tool that plays a vital role in the SPIRIT SoC automation flow. Recall that the main objective of the

SPIRIT standard is to automate the SoC design flow from the XML design description to an operational simulation or implementation. The netlister can therefore be considered as the most important automation tool.

The standard input for the SoC simulation are formal sources as RTL or SystemC, which can be compiled and executed by a simulation kernel. Be it any level of abstraction, a top netlist is required to perform the platform simulation, e.g. an RTL/TLM top netlist is necessary for simulating a mixed RTL/TLM design. The netlist is a purely structural description that is essentially a list of component instances with interconnections between their interfaces. Note that no algorithmic or behavioral codes should be included in a netlist.

For any given platform, either an RTL or SystemC netlist is required for its simulation. Thus, two types of netlisters are available in general:
1. RTL netlister to generate the RTL netlist;
2. SystemC netlister to generate the SystemC netlist (for TLM).

3.5.1   Co-Simulation Netlist

The co-simulation is typically a simulation that mixes both TLM and RTL models. It is needed for simulating a complex RTL design with a huge number of elements to support high-level functionalities. Recall that this is indeed the reason of developing the TLM methodology to represent the behavior of an RTL design with only the algorithm using a system-level language.

The users will construct a mixed test-bench by choosing the appropriate IPs to be simulated at RTL as the Design Under Test (DUT), and those to be simulated at TLM for its behavior from the system point of view. Once the choices are made, the netlister tool will provide the corresponding netlist automatically.

3.5.2   Co-Emulation Netlist

To increase the execution speed of a simulation, certain synthesizable RTL blocks can be mapped on emulators. A different netlist than the co-simulation netlist must be generated by the netlister tool.

3.6   Other Generators

Many other kinds of generators can be created according to what the users need to do. The only condition to create a generator is that the necessary information must be written in the XML document.

3.6.1   Regression Generator

Many of the SoC platform IPs especially the host processor need to execute the embedded software. The role and the amount of the software has become increasingly significant in the SoC design. Therefore, it is very useful to describe a list of test codes to execute an IP for a full regression test. Such description can be written in an XML document, based on which the regression test suite can be generated.

3.6.2   Register Access Test Generator

In the SPIRIT schema, the memory map of an IP can be described with the accurate descriptions of all of its registers and register fields. Specific generators can generate the SystemC header files for the IP, which contains the definitions of all registers. These generators can also generate the software that will be executed on the simulation platform. This software will try to access the IP registers to verify if the access rules are well respected.

4.   EXAMPLE

4.1   Platform Architecture

The TC4SoC[7] platform serves as a demonstrator to validate the design automation strategy described earlier on a real platform. TC4SoC is a test chip vehicle that validates several IP blocks and CAD tools in 90nm CMOS technology. It is a SoC design comprising PCI and LMI interfaces. It provides the STBus External Port (SEP) that enables the interconnection to external high-speed buses. This chip supports flash memories and a board range of peripherals connected via a programmable glue logic. Other memories included are embedded ROM, SRAM, and eDRAM. IPs are interconnected through the STBus interconnect, which contains four STBus nodes as transaction routers. Figure 7-8 illustrates the structure of TC4SoC.

---

[7]   A SoC design developed by STMicroelectronics.

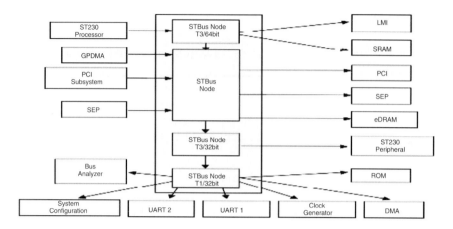

Figure 7-8. The Architecture of TC4SoC Platform

4.2   IP Packaging and Platform Generation110

This section describes the IP packaging and the platform generation following the SPIRIT strategy. The UART IP in the TC4SoC platform will be given as an example of the SPIRIT compliant component.

To begin with, a SPIRIT component file must be created for the UART. The RTL entity (VHDL in this example) of the UART is used as the entry point to create this component file. The signal section under the *hwModel* section of the component file corresponds to the signal list of the RTL entity. This mapping can be done manually or automatically by a tool, *vhdl2spirit*.

If signals are correctly named in the RTL entity, for instance, giving the same prefix to a group of signals belonging to the same bus interface and using the standard names to indicate bus types, then the *vhdl2spirit* tool is able to detect bus interface type and create the *busInterface* section with the corresponding signal mapping.

As depicted in Figure 7-9, the example of UART component has three bus interfaces, i.e. a slave STBus T1 interface, an input clock interface, and an input reset interface. There are three types of STBus signals: T1, T2, and T3. Since the STBus interface used in the UART is only T1, all of the STBus signals are prefixed with *stbus1* in the UART VHDL entity. As a result, the *vhdl2spirit* tool can detect correctly the STBus interface. The script can also detect automatically that the interface is T1 since T2/T3 signals are missing.

After instantiating all the design IPs, users are now ready to connect them to the STBus interconnect. The STBus interconnect is a standard yet fully configurable IP. For instance, the number of bus interfaces is not static as it depends on the number of IPs connected to it. It is therefore impractical to store all of the possible interconnections in a database.

The XML database contains a template of the STBus interconnect. This template is processed by a style sheet in the eXtensible Stylesheet Language (XSL) to generate an XML file with the appropriate number of bus interfaces.

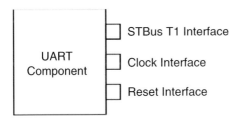

Figure 7-9. Bus Interfaces of UART in TC4SoC Platform

The remaining tasks include the configuration of signal size and the pin connection between master and slave interfaces, which are performed by a SPIRIT generator. First, the SPIRIT design environment generates a Loose Generator Interface (LGI) file that describes the environment. This file is indeed the input file to help the generator to locate the paths to different XML files.

When the design and all component instances are loaded, the configuration generator will search for the STBus interconnect instances. Once the interconnect instances are identified, the configuration generator will loop through the STBus interfaces of these instances. For each interface, the generator will try to match its signal sizes to the connecting interface with respect to the STBus specification. Once matched, the interfaces are interconnected. When the interconnection task is completed, the generator will check and remove any unused signals in the STBus interfaces. A new XML file is then generated with the information of the new path for the interconnections. This file is passed to the netlister in the form of the Loose Generator Change (LGC) file.

Tasks accomplished up to this point include the platform configuration, the IP configuration, and the connections of the design-level bus interfaces. The netlister tool can now perform its job with all the available information to generate a top-level RTL netlist of the TC4SoC platform for simulation. Figure 7-10 shows the SPIRIT design automation flow.

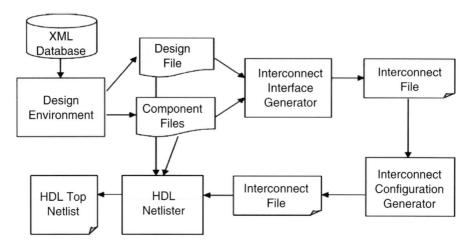

Figure 7-10. SPIRIT Design Automation Flow

REFERENCES

[1]   SPIRIT Consortium web site available at: http://www.spiritconsortium.org

[2]   SPIRIT Schema Working Group, "Spirit User Guide V1.0", December 2004.

Abbreviation

API	Application Programming Interface
ASIP	Application Specific Processor
ATPG	Automatic Test Pattern Generation
BCA	Bus Cycle Accurate
CMOS	Complementary Metal Oxide Semiconductor
CODEC	Coder/Decoder
CPU	Central Processing Unit
DCT	Discrete Cosine Transformation
DMA	Direct Memory Access
DOM	Document Object Model
DSP	Digital signal processing
DUT	Design Under Test
EDA	Electronic Design Automation
eDRAM	Embedded Dynamic Random Access Memory
FIFO	First-In-First-Out
FPGA	Field Programmable Gate Array
FSM	Finite State Machine
GALS	Globally Asynchronous Locally Synchronous
GDB	GNU DeBugger
GDS	Graphical Data System
HDL	Hardware Description Language
ICE	In Circuit Emulator
IP	Intellectual Property
IPTG	Intellectual Property Traffic Generator
ISS	Instruction Set Simulator
JTAG	Joint Test Action Group

LCD	Liquid Crystal Display
LGI	Loose Generator Interface
LMI	Local Memory Interface
MDA	Model Driven Architecture
MMU	Memory Management Unit
NoC	Network on Chip
NUMA	Non-Uniform Memory Architecture
OS	Operating System
OSCI	Open SystemC Initiative
PC	Program Counter
PCI	Peripheral Component Interconnect
POSIX	Portable Operating System for unIX
RDB	Relational Data Base
RISC	Reduced Instruction Set Computer
ROM	Read Only Memory
RPC	Remote Procedure Call
RTL	Register Transfer Level
SAT	Satisfiability problem
SCE-MI	Standard Co-Emulation Modeling Interface
SCV	SystemC Verification
SDI	Streaming Data Interface
SEP	STBus External Port
SMP	Symmetric Multiprocessor Servers
SoC	System-on-Chip
SPAG	SysProbe Analysis Generator
SPES	SysProbe Embedded Software
SPIRIT	Structure for Packaging, Integrating and Re-using IP within Tool-flows
SQL	Simple Query Language
SRAM	Static Random Access Memory
TLM	Transaction Level Modeling
UART	Universal Asynchronous Receiver and Transceiver
UML	Unified Modeling Language
VLIW	Very Long Instruction Word
VLNV	Vendor Library Name Version
XML	eXtensible Markup Language
XSD	XML Schema Definition
XSL	Extensible Stylesheet Language

Index